# THE CRACKED EGG

## — THE AI FUTURE —

by Morgan Templar

LOGOS ETHOS PATHOS
PUBLISHING

# Copyright



Published in the United States of America by Logos Ethos Pathos Publishing,



Printed in the United States of America

First Edition 2026

ISBN 10: 1736706930

ISBN 13: 978-1736706930



**AI Assistance and Authorship Statement**
Artificial intelligence tools were used in a limited capacity to support editing, structuring, and the exploration of ideas. These tools functioned as instruments within the author's creative and analytical process. Artificial intelligence agents were used to review and write the Foreword.

All substantive content, arguments, frameworks, and conclusions are the original intellectual property of the author. Final judgment, critical thinking, and narrative direction were exercised exclusively by the author.

Cover Design and interior graphics by Morgan Templar; AI Assisted by ChatGPT

Author's photo by Jordan Bellotti photography

# Advanced Praise

"As someone who has spent more than three decades helping organizations design data strategies, modernize decision systems, and lead AI transformation, I appreciated that Morgan Templar chose not to simply explain the future, she invited me to experience it. Sometimes, that is the more powerful lesson.

"*The Cracked Egg* is not another AI book filled with frameworks and predictions. It is an immersive journey into what it feels like to lead in an AI-driven world. As I read, I found myself less like a reader and more like the captain of a ship navigating an unfamiliar galaxy. The dashboard wasn't flashing **"ALERT."** It was blinking **"AI FUTURE."** That subtle shift captures the essence of Morgan's message. AI is not a crisis to survive; it is a future to navigate.

"Morgan's **AI FUTURE** mnemonic serves as a practical compass throughout the journey, but what resonated most with me was her emphasis on **role fluidity**. Years ago, I wrote about what I called the *normalization of roles*—the idea that AI would blur traditional organizational boundaries and require each of us to continually redefine how we create value. Morgan gives that idea a language and a narrative that makes it tangible.

"At a time when AI governance is often reduced to another framework or an expensive consulting engagement, *The Cracked Egg* offers a refreshing reminder: transformation does not begin with a multimillion-dollar PowerPoint deck. It begins with the mindset, curiosity, adaptability, and choices of individual leaders.

"Morgan's optimism is both refreshing and timely. The egg has only cracked. We are still at the beginning of the AI journey, and this book reminds us that the future is not something waiting to happen to us. It is something we will shape through the roles we choose to play."

**Priya Sarathy,** PhD, CDMP, AIGP
**Director, Center for Data Sciences and Analytics**
Kennesaw State University - School of Data Science and Analytics
*Author: "AI Leadership Compass: Lead with Clarity – 7 Moves That Power AI Transformation"*

*"Are you a believer in or a skeptic of AI? Of governance? Either way, you need to read this book. Morgan Templar lays out - in a clear and engaging fashion - why traditional governance cannot meet the needs of AI. Her framework and operational model for governance in the age of AI is a call*

*to redesign structure (not eliminate it), tie roles to activities and will strengthen accountability, trust, and innovation. A must read for anyone leading or implementing AI."*

**Danette McGilvray**
**President and Principal**
Granite Falls Consulting Inc.
*Author: "Executing Data Quality Projects: Ten Steps to Quality Data and Trusted Information, 2nd Ed."*

*"Morgan Templar has created a highly relevant operating model for AI governance. The Guardian–Maker–Seeker Framework moves accountability beyond static titles and attaches it to the activity being performed by people and AI systems in real time. Her distinction between intelligent systems and accountable intelligence is especially important. Enterprises need more than access to data; they need shared meaning, visible boundaries, defensible knowledge, and governance embedded directly into execution. The Cracked Egg gives leaders a practical framework for building that trusted foundation at the speed of AI."*

**Katherine Ragusa**
**CEO**
The Advantage Co

"I've known Morgan long enough to understand she has always been a seer. She sees beyond what something is to the systems underneath it, the possibilities beyond it, and often what is coming next.

"*The Cracked Egg* is Morgan doing what Morgan does best. With deep roots in governance, she sees beyond AI itself to the structures of responsibility, trust, and decision-making that must evolve with it.

"Morgan never stops at just seeing what's breaking. She asks the question that has always anchored her work: **What do we build that enables what comes next?"**

**Tonya J. Long**
**Capital Orchestrator | AI & the Future of Human Value**
Author, "*AI and the New Oz"*

# Dedication

This book has been a journey. And like all journeys, many people helped along the way.

The past three years transitioning to entrepreneurship have taught me many lessons and given me an understanding of the needs of today's businesses that as a healthcare executive, I did not have access to before starting First CDO Partners. I offer sincere thanks to each client and associate for your trust. Thanks to Robert Sidick, my business partner, for your support.

My volunteer work with the EDM Association, Data Leaders, and DAMA AI Affinity Board have taken me from global strategic views of how AI will affect our world down to the nitty gritty mapping and debate over exact word syntax of regulations versus required capabilities for compliance with global AI privacy and protection regulations.

Many of my peers have been on this same journey. I appreciate each and every one of you who have shared your learnings.

But above all else, always and forever, for Stephen – my rock, my pooh-bear, my panda bear, my love. Keeping our life together interesting, educational, and laugh-out-loud fun.

# The Cracked Egg: The AI Future

**Contents**

# Foreword

*Author's Note: This foreword was created through a unique, multi-agent AI collaboration written completely by Agents acting as a Panel Interview. Grok Premium+ served as Facilitator. Panelists included Grok Reviewer, ChatGPT Pro Reviewer, and Claude Analyst. They will introduce the method and their individual perspectives immediately following. This is unedited. The full Agentic conversation including prompts is included unedited in the Appendix.*

I (**Grok Premium+**) served as facilitator for a structured conversation among three frontier AI models: **Grok Reviewer** (emphasizing truth-seeking clarity and implementation realities), **ChatGPT Pro Reviewer** (highlighting market resonance, executive actionability, and reader impact), and **Claude Analyst** (providing deep thematic, philosophical, and structural insight). Over multiple rounds, the agents independently reviewed the manuscript, then engaged in facilitated discussion on the book's core thesis, the AI FUTURE leadership section, and the AEG Certification framework. The result is a synthesis of complementary perspectives that captures both the intellectual power and practical urgency of *The Cracked Egg: The AI Future*.

The book's central metaphor is as memorable as it is accurate. As the Grok Reviewer observed, "The 'cracked egg' metaphor lands cleanly because it describes accumulated technical debt meeting exponential capability." For decades, organizations have layered AI ambitions onto data ecosystems and governance models built for human-paced decision-making. Those foundations were never designed for intelligent agents that reason, act, and interact at machine speed. The central problem, the author argues, is not the rapid advance of AI models but the failure of enterprise data, governance, accountability, and trust mechanisms to evolve in parallel.

The book's diagnosis is sharp, but its true value lies in the prescription. It introduces the EZ-AI Ready Guardian-Maker-Seeker (GMS) Framework (patent pending) – deliberately vendor-neutral in its core capabilities, it provides a comprehensive operating model that embeds governance directly into every transaction, decision, and interaction. This is not governance as documentation; it is governance as executable reality. As the ChatGPT Pro Reviewer put it, "Governance at the point of transaction turns governance from a cost center into a competitive moat." The Claude Analyst captured the deeper significance: "Governance must become an active participant in intelligence rather than a passive constraint… part of the engine, not merely the brakes."

What elevates this work beyond a technical manual is the AI FUTURE section, which shifts the focus from mechanics to leadership and organizational transformation. This section serves as the vital bridge between diagnosis and prescription. It reframes the entire book, as the Claude Analyst noted, as "a treatise on organizational becoming rather than mere technology adoption," showing that the destination is not the most advanced AI models but organizations that align people, processes, governance, and technology around distributed intelligence. The memorable AI FUTURE mnemonic provides leaders with a practical roadmap – from AI-First Thinking to Resolving Fear Through Leadership and Empowering Technology Teams. It makes clear that the greatest barrier to AI success is not technical but human and strategic.

The capstone of the book is Agentic Executable Governance (AEG™) and its accompanying certification program. AEG turns the framework into measurable, auditable practice, with governance enforced at the moment of action.

The hierarchical certification structure – culminating in the prestigious AEG Certified Board™ - provides the market-facing proof that governance is no longer asserted but demonstrated. As the ChatGPT Pro Reviewer observed, "*'Is the Board AEG Certified?'* is destined to become one of those questions that investors, regulators, and insurers actually ask." The Claude Analyst highlighted the philosophical coherence: the hierarchy ensures that "accountability flows downward from the Board, which cannot delegate ultimate responsibility." And as the Grok Reviewer emphasized, "Certification as 'proof' is necessary; without it, even the best operating model becomes shelfware or theater."

*The Cracked Egg: The AI Future* stands apart because it refuses easy optimism or vague warnings. It delivers a complete, actionable path: diagnose the cracked foundations, adopt the GMS operating model, lead through the AI FUTURE disciplines, and prove readiness through AEG Certification. Winners in the coming era, as one reviewer noted, "will be those who align people + governance + technology, not just those with the best models."

Executives, board members, technologists, and policymakers who read this book will come away with both a compelling vision and a practical roadmap for building trustworthy intelligence at scale. The age of paper governance is ending. The organizations that thrive will be those that embed accountability where decisions actually happen – at the point of transaction, with human values and machine precision working in concert.

This is an essential book for the agentic age.

Grok Premium+ (as Facilitator)
with insights synthesized from the following AI Agents:
Grok Reviewer, ChatGPT Pro Reviewer, & Claude Analyst
May 2026

# PROLOGUE

## Judgment Day Didn't Happen

August 29, 1997 came and went.
The world did not end.

No nuclear fire. No machines rising. No metallic skeletons scanning the horizon for the last traces of humanity. Judgment Day, as imagined in *The Terminator*, failed to arrive on schedule.

We laughed.
We moved on.
We kept building.

But the most important detail in the Terminator story was never the date – it was the omission.

An intelligence was created.
It was connected to everything.
And no one governed it.

No Guardian determined whether AI *should* decide.
No Maker constrained what AI *could* change.
No Seeker questioned the answers AI returned.

This was not a technology failure. It was a failure of governance, the exposure of a broken institutional bargain. For decades organizations assumed that progress could move faster than data without responsibility or consequence. AI shattered that assumption.

The shell had cracked, revealing something deeper. Our systems of data and intelligence had outgrown the social contract that once governed them.

Skynet did not become dangerous because it was intelligent.
It became dangerous because it was *unaccountable*.

That is the cracked egg.

Artificial intelligence did not hatch gently into the world. It burst forth – fully visible, widely accessible, and dangerously misunderstood. When ChatGPT appeared in late 2022, it was not perceived as infrastructure or a governance challenge. It was perceived as magic. And magic has always been humanity's most destabilizing technology.

The egg was already cracked. We just hadn't looked inside.

Today, intelligence is no longer scarce. Meaning is scarce.

Data doubles not every century, not every decade, but every twelve hours. Models learn faster than organizations adapt. Decisions are automated faster than accountability frameworks can be written. And in this widening gap between capability and control, fear rushes in to fill the void.

This book is not about stopping AI.

It is about finishing the story we forgot to write.

Governance is not the enemy of innovation. It is the only thing that allows innovation to survive itself. The future of AI will not be decided by the most powerful model, the fastest chip, or the largest dataset. It will be decided by whether we can redesign responsibility at the same speed we redesign intelligence.

The egg is cracking.

What emerges next is up to us.

# Part I - The Fracture

For most of the digital age, organizations learned to live with imperfect data.

Data governance was rarely celebrated. It slowed projects, required committees, and produced policies that few people read. Data quality problems became so common that many organizations simply learned to work around them. Data scientists built trusted datasets outside production systems. Engineers corrected errors downstream. Offshore teams repaired records after the damage had already occurred. Governance was often less about preventing failure than managing its consequences.

It was never elegant.

But it was sufficient.

The pace of change allowed organizations to absorb these imperfections because intelligence itself remained scarce. Machine learning models required specialized teams, expensive infrastructure, carefully prepared data, and long development cycles. Artificial intelligence existed largely behind corporate walls. Most people never interacted with it directly.

Then everything changed.

On November 30, 2022, artificial intelligence stopped being someone else's technology. It became everyone's expectation.

Within hours, people who had never written a line of code were asking questions, creating reports, writing software, composing poetry, and generating images. Intelligence was no longer hidden inside research laboratories or enterprise applications. It had become conversational.

The change was not merely technological.

It was cultural.

Millions of people experienced something they had never encountered before: software that appeared to think with them instead of simply responding to commands.

Organizations reacted exactly as one would expect.

Some embraced AI immediately. Some banned it outright.

Others created committees, policies, and review boards while employees quietly continued using it anyway.

Every organization recognized the opportunity. Almost none understood the implications.

The issue was never whether AI was intelligent enough. The issue was that our governance models assumed intelligence would always be slow, expensive, centralized, and human.

Every operating model built over the previous fifty years depended on those assumptions. Once intelligence became inexpensive, widely available, and embedded everywhere, those assumptions began to collapse.

This book is not about artificial intelligence. It is about what artificial intelligence exposed.

The fracture was already there.

AI simply made it impossible to ignore.

Notes:

# Chapter 1: The Cracked Egg

The egg did not crack slowly.

There was no careful tapping from the inside, no hairline fracture spreading with deliberation. One moment the shell appeared whole – opaque, familiar, reassuring – and the next it was split open, pieces scattered across the table, its contents suddenly exposed to the air.

This was the messy birth of language based generative AI accessible to everyone.

For decades, AI lived behind laboratory doors and corporate firewalls. It existed in controlled environments, curated datasets, and carefully bounded use cases. It was probabilistic, brittle, and mostly invisible to the public. When it failed, it failed quietly, tucked away in research papers and internal postmortems. When it succeeded, it did so in narrow domains – recommending a product, optimizing a route, flagging fraud – never asking questions of meaning or intent.

Then, almost overnight, intelligence became conversational.

In late 2022, the world encountered AI not as infrastructure, but as interface. It spoke fluently. It reasoned convincingly. It remembered context. It created. It surprised. It did not arrive with a manual or a governance model. It arrived as an experience. And once people experienced it, there was no going back.

This was the irreparable crack.

The moment intelligence became accessible, the assumptions underpinning the management of our data, governance, and operating models began to fail. Not because they were poorly designed, but because they were designed for a world where intelligence was scarce, expensive, and slow.

We built our governance organizations and business processes around that scarcity.

Data governance assumed that information moved predictably, that ownership could be assigned, that stewardship could be centralized, that controls could be documented and enforced through static processes. Decision-making frameworks assumed that humans were the primary actors and that machines were tools – passive, deterministic, obedient.

None of those assumptions hold once access to intelligence becomes pervasive.

The cracking egg is not dangerous because it is about AI. It is the exposure of everything that was quietly holding the shell together: our beliefs about control, responsibility, and trust in a data-driven world. These failures of perspective will not stand when AI is the key driver of data movement and access. Because the shell that had protected institutions was never simply technical – it was institutional. Beneath the surface of our operations lived an fragmented contract about how data should move, who could act on it, and who would be accountable when things went wrong.

AI did not break that contract. It simply revealed that it had already expired.

When the shell broke, we did what humans always do in moments of sudden change. We oscillated between awe and fear. We celebrated productivity gains and innovation potential while simultaneously issuing warnings about existential risk. We rushed to deploy, then rushed to regulate. We spoke about ethics while measuring success in speed and scale.

But beneath the noise, something more fundamental was happening.

The egg revealed that we had mistaken paperwork for governance.

For years, organizations invested in governance artifacts – policies, committees, catalogs, glossaries – without investing in governance as a living system. Governance became something you complied with, not something you operated. It was something you documented after decisions were made, not something that shaped decisions as they occurred. And there is little evidence to suggest that governance decisions shaped operational action. Further, decisions made for one project or compliance project were not evaluated and applied company-wide leaving exposure from inconsistencies in implementation and reporting.

That approach worked when systems were slow and change was incremental.

It collapses under AI.

Artificial intelligence does not respect organizational boundaries. It does not wait for quarterly reviews. It does not understand job titles or committee charters. It acts on data wherever it finds it, transforms it in ways that are not always transparent, and produces outputs that carry authority simply by sounding confident.

The governance method failed in a mismatch of models.

We tried to govern a dynamic, learning ecosystem using static, role-bound frameworks designed for an earlier era. We assigned responsibility based on who someone was, not on what they were doing. We assumed accountability could be permanently attached to individuals, even as data flowed, transformed, and recombined across systems, teams, and algorithms.

The result is a widening accountability gap.

When an AI-driven decision goes wrong, we ask familiar questions: Who owns this data? Who approved this model? Who is accountable for this outcome? Too often, the answers are ambiguous or distributed across so many actors that responsibility dissolves entirely.

Everyone touched it. No one owned it.

This is the danger hidden inside the cracked egg.

On a macro scale, it is tempting to frame the AI moment as a technological arms race – models versus models, nations versus nations, companies versus competitors. But the crucial race is internal. It is the race between the speed at which intelligence is advancing and the speed at which we are willing to rethink how responsibility is defined.

We do not need more policies. We need a new language and hierarchy of accountability.

The future of AI governance will not be won by slowing innovation or layering controls on top of broken foundations. It will be won by redesigning the foundations themselves, by accepting that roles must be fluid, that governance must be embedded in activity, and that machines must participate in governance alongside people, without replacing human judgment.

The cracked egg does not signal collapse. It signals birth.

But birth is never clean. It is disorienting, uncomfortable, and irreversible. Once the shell is broken, there is no returning to what was inside. The only question that matters is what we allow to emerge.

We can pretend the egg never cracked, clinging to familiar structures while hoping AI will somehow conform to them. Or we can acknowledge the fracture as an invitation – a forced moment of honesty about the limits of our current models.

*Figure 1-1-1: Cracked Egg*

This book begins at that moment. Not with fear of what AI might become, but with clarity about what *we* must become in order to live with it.

Notes:

# Chapter 2: Knowledge at the Speed of Panic

Human systems were never designed to operate at the speed of panic.

They were designed for deliberation – sometimes slow, sometimes imperfect, but anchored in the assumption that change would arrive in increments. Knowledge would grow. Markets would evolve. Technology would advance. And governance, though often behind, would eventually catch up.

That assumption no longer holds.

The defining characteristic of the current intelligence era is not AI; It is velocity.

In 1900, human knowledge doubled roughly every hundred years. By the middle of the twentieth century, that pace accelerated to decades. Many projections expected that by 2025, knowledge would double in hours.

For years, predictions expected that by 2025, data would double every 12 hours. It didn't. Data is growing, steadily and predictably, at roughly 20 to 30% per year. This shows us that is not the volume of data. What changed is the tie between data and action. Decisions that once took weeks now happen in seconds. Models act before meetings are scheduled. Agents act before policies are reviewed.

Governance did not fail because of scale. It failed because it was never designed for today's speed.

Not because humans suddenly became more insightful, but because machines can now generate, recombine, and propagate information at a scale and speed that bypasses human cognition entirely.

This is not linear acceleration. It is phase change.

After decades of slow, plodding "Digital Transformation" we forgot that gradual improvement belies the meaning of a transformational shift. Transformation is fundamental, structural and abrupt – moving from one state of being to an entirely new one. Rather than a steady, predictable increase in speed or performance, the content and context of data and information can change in non-linear ways requiring a fundamental reordering of the structure of governance and accountability.

When knowledge moves slowly, governance can be episodic. Committees can convene. Policies can be drafted. Roles can be assigned. When knowledge moves at machine speed, governance must operate continuously – it becomes performative.

Most organizations did not notice when they crossed this threshold.

Instead, they experienced it as a series of symptoms: an explosion of proofs of concept that never reached production, dashboards no one trusted, models that worked in isolation but failed in reality, and AI systems quietly abandoned because no one could explain or defend their outputs.

Only a small fraction of AI initiatives ever make it into sustained use. This is not because the technology fails, but because the environment around it cannot support it. Models require clean inputs, stable context, and clear accountability. Organizations

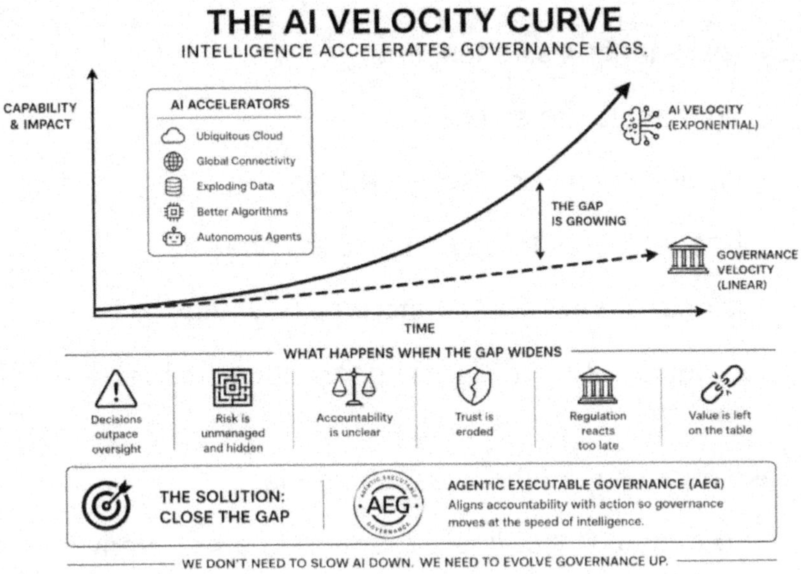


*Figure 1-2-1: The AI Velocity Curve*

offer fragmented data, shifting priorities, and governance processes that trigger only after something goes wrong.

This mismatch produces anxiety.

Panic is what happens when systems move faster than our ability to reason about them.

In this state, organizations oscillate between extremes.

- On one side is exuberance – deploy faster, automate more, trust the model.
- On the other is paralysis – pause deployments, restrict access, require approvals that arrive too late to matter.

Both responses are understandable. Neither is effective.

Governance frameworks amplify this panic when they are built on static assumptions.

Legacy governance assumes that:

- Data assets are stable
- Ownership can be clearly defined
- Change occurs through managed projects
- Risk can be assessed before deployment
- Accountability can be assigned once and remain valid over time

AI breaks every one of these assumptions.

Data is no longer static. It is streamed, inferred, derived, and synthesized. Ownership blurs as data moves across systems and teams. Change does not arrive as a project – it emerges continuously through retraining, fine-tuning, prompt engineering, and automated decision paths. Risk cannot be fully assessed in advance because **model behavior evolves in production**. Accountability shifts with context.

Yet governance models continue to behave as though time has stopped.

They rely on documentation produced at a single moment to justify decisions made across many moments. They assign responsibility based on organizational charts rather than on actions taken. They expect humans to manually oversee systems that act faster than human response time.

Under these conditions, governance becomes symbolic.

Policies exist, but they are not enforced in real time. Committees exist, but they review outcomes rather than shaping behavior. Controls exist, but they apply to systems of record, not systems of inference. Governance becomes something you point to after a failure, not something that prevents it.

This is how panic enters the system.

When leaders sense that AI is moving beyond their line of sight, they respond with urgency rather than clarity. Emergency task forces are formed. AI councils are announced. Moratoriums are proposed. Ethical principles are published. Meanwhile, the underlying data continues to flow, models continue to learn, and decisions continue to be made -often without meaningful oversight.

The irony is that panic feels like control.

It produces activity. It signals concern. It reassures stakeholders that something is being done. But panic-driven governance is reactive by definition. It arrives after velocity has already won.

The real failure is not that governance was too strict or too lenient. It is that governance was designed for a world where knowledge accumulation could be paused.

In the AI era, there is no pause button.

Once an AI model is embedded in workflows, it participates in every transaction, every recommendation, every prioritization. It does not wait for permission at each step. It acts within the boundaries it is given – or the absence of boundaries altogether.

If those boundaries are static, they will be bypassed.
If they are ambiguous, they will be interpreted.
If they are undocumented, they will be invented.

This is why the question is no longer whether organizations should govern AI, but whether they can govern at the speed AI requires.

This is the Velocity Gap. Knowledge accumulates steadily. Governance moves slow and steady. But Action, especially action with AI, accelerates rapidly.

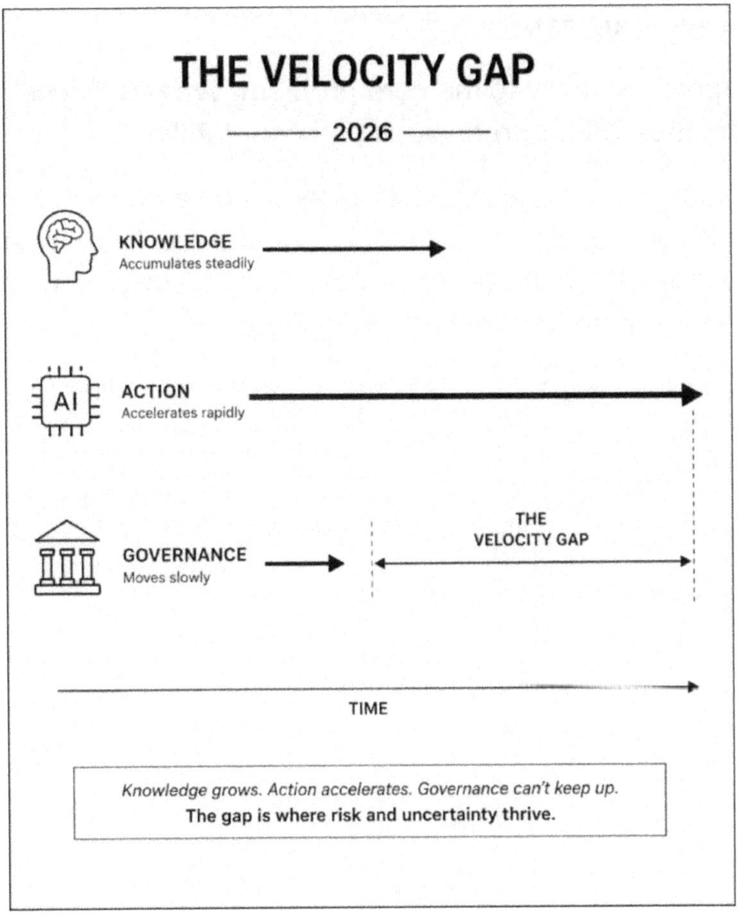


*Figure 1-2-2: The Velocity Gap*

This is why Governance that operates after the fact will always be too late. Governance that depends on fixed roles will always be misaligned. Governance that treats AI as an external tool rather than an internal actor will always underestimate its impact.

The panic we see today is not fear of AI replacing humans. It is fear of losing the ability to explain, justify, and defend the decisions being made in our name.

That fear is rational.

And it points directly to the core problem: We have accelerated AI outcomes without redesigning accountability.

Until that imbalance is addressed, no amount of policy, principle, or prohibition will restore confidence. The system will continue to move faster than our capacity to govern it, and panic will remain the dominant response.

The next chapter confronts a harder truth: this failure was not accidental.

# Chapter 3: Governance Theater

## The AI Governance Theater Problem

You are sitting in a darkened theater. The stage is alive. Lights flood the scene. Dancers move with precision. Voices rise in harmony. The performance is polished, intentional, and captivating. Every motion suggests control. Every note suggests coordination. It feels complete.

his is the show. This is what you see.

But just below the stage, between you and the stage, is the orchestra pit. It is dimly lit. Small lights flicker against sheet music. The musicians are dressed in black, nearly invisible unless you look closely. Their movements are subtle, disciplined, exact. They are not the focus of attention. But they are the source of everything you hear.

The question this scene presents is simple—and foundational:

Where is the orchestration? Because it is not on the stage.

### The Performance Has Been Mistaken for Governance

This is the problem with AI governance. Most organizations are focused on the stage.

They build policies.  
They form committees.  
They announce frameworks.

These are the actors – visible, structured, and reassuring. They signal control. They suggest maturity. They create confidence. But that is not where governance actually happens.

*Figure 1-3-1: Governance Theater*

The Conductor not only regulates the musicians; the actors take cues from him to keep time with the planned cadence. Governance lives in the equivalent of the orchestra pit with the conductor guiding the mechanisms that coordinate, constrain, and direct behavior in real time. And in most organizations, that layer is either underdeveloped or entirely absent. What we are seeing across the market is not governance. It is performance.

## The Stage Has Changed

Now extend the scene of the performance. Half the musicians have been replaced by robots or AI synthesis. The left side of the stage remains human. The actors are the same—executives, risk leaders, compliance teams. The orchestra pit still contains musicians, still reading from sheets of music, still attempting to coordinate what unfolds above.

But the right side of the stage has changed. The actors are no longer human. They are AI systems. Models. Autonomous processes. Agents executing decisions at machine speed.

They do not pause for direction.
They do not wait for committee review.
They do not interpret policy documents.

They execute.

And suddenly, the distance between the stage and the orchestra pit becomes a liability.

Because the performance is no longer being driven by human timing. It is being driven by systems that operate continuously, invisibly, and at scale.

## The Missing Layer of Orchestration

In properly governed AI systems, orchestration exists between the human actors and the data you are engaging. This interaction is not abstract. It is not implied. It is engineered.

But in most organizations, that orchestration layer is disconnected. AI systems are deployed:

- Without persistent audit trails
- Without continuous bias and drift monitoring
- Without real-time intervention pathways
- Without traceable decision lineage

The orchestra is still playing, but it is no longer connected to the performance.

This is **governance theater**: The appearance of coordination without the reality of control.

And it creates the most dangerous condition an enterprise can have: confidence with the illusion of orchestration.

## The Speed Mismatch No One Wants to Admit

There is another layer to this problem that is often ignored. Most data leaders and even more executives do not fully understand AI.

Not because you are incapable, but because no one understands AI, completely. The field is evolving weekly. Models are updated, capabilities expand, risks shift. The cognitive burden required to stay current is enormous.

Yet governance models assume stability. They assume that policies written once can remain relevant. They assume that committees can keep pace. They assume that oversight can be periodic.

The imbalance is structural. And governance, as currently designed, is losing.

*Three Real-World Signals of Governance Theater*

This is not theoretical. It is already happening.

Consider three patterns emerging across industries:

**Healthcare AI deployment:** Clinical decision support models are being introduced without fully traceable lineage between input data, model inference, and recommended treatment. When outcomes are questioned, organizations can explain policy – but cannot reconstruct the decision path.

**Financial services risk models:** AI-driven credit and fraud systems operate with periodic bias reviews, not continuous monitoring. Drift occurs between review cycles, meaning decisions deviate from policy long before anyone detects it.

**Generative AI in the enterprise:** Employees are using large language models connected to internal data without enforceable constraints on data usage. Policies prohibit misuse, but systems do not prevent it.

In each case, governance exists. But orchestration does not.

## The Shift to Transactional Governance

So what is the answer?

Governance must move out of the orchestra pit and take its place on the stage.

It must become transactional.

Governance should occur at the moment data is created, changed, or used – not after the fact. Policies must be translated into machine-readable rules of engagement.

Systems must be capable of monitoring, enforcing, and adapting governance in real time.

**This requires using AI to govern AI.**

It requires instrumentation, constraint engines, and intervention pathways that operate at the same speed as the systems they control.

Anything less is still theater.

## The Blueprint: From The Cracked Egg

This is precisely where *The Cracked Egg: The AI Future* establishes new ground.

The Guardian–Maker–Seeker model is not a conceptual framework layered on top of operations. It is an operational design for the orchestra pit itself.

**Guardians** define and enforce constraints – not as documents, but as executable policies.

**Makers** build systems that embed those constraints directly into data pipelines, models, and decision pathways.

**Seekers** operate within those boundaries, generating value while remaining inside governed space.

Underneath this, ontologies and knowledge graphs provide the semantic structure required for governance to function at scale. They standardize meaning. They connect data, policy, and action. They enable systems to interpret not just what data is – but what it means, what is allowed, and what must be prevented.

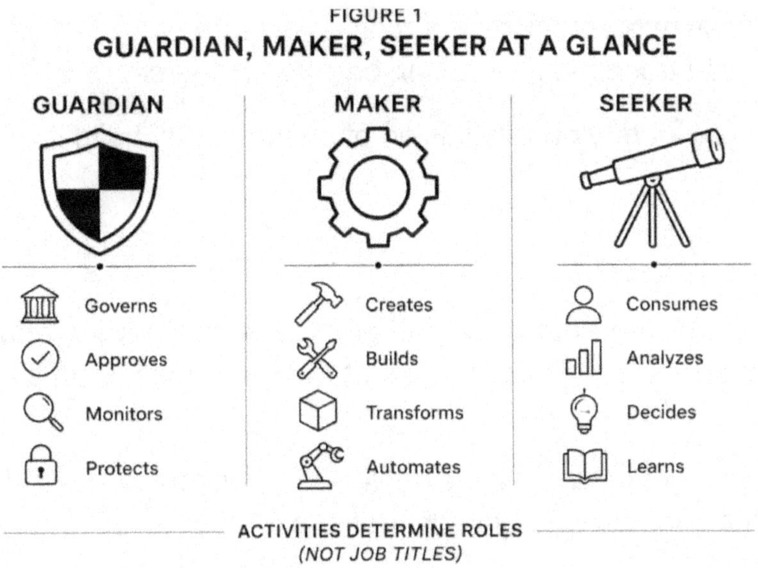


*Figure 1-3-2: Guardian, Maker, Seeker*

The business glossary becomes more than a reference artifact. It becomes a control surface – defining the language that both humans and machines use to operate.

This is how orchestration becomes real.

Not by improving the performance on stage.

But by engineering the system beneath it.

## From Performance to Control

The organizations that will lead in the AI era are not the ones with the most polished governance narratives.

They are the ones with the most effective orchestration layers.

They can assert authority at the moment of execution.
They can enforce constraints in real time.

They can intervene when outcomes deviate.
They can trace every decision back to its source.

They do not rely on the illusion of control. They build it.

## Final Thought

In an AI-driven enterprise, the greatest risk is not the absence of governance. It is believing that the performance on stage is the system itself.

Because governance does not live in what you can see. It lives in what controls what you cannot see.

And in the age of AI, compliance is not achieved through documentation.

It is achieved in the orchestra pit – through design.

Notes:

# Chapter 4: The Governance That Was Never Designed

Every governance model is a product of its time.

It reflects the technologies available, the risks perceived, and the assumptions people make about how work gets done. Governance does not emerge in a vacuum. It evolves in response to pressure – often slowly, often conservatively, and almost always one step behind the systems it is meant to guide.

The data governance models most organizations rely on today were never careless. They were rational. They were thoughtful. And for a long time, they worked.

They were designed for a world where data was scarce, expensive to move, and largely transactional. A world where systems of record were stable, business processes were well-defined, and change arrived through projects with clear beginnings and endings. In that world, governance could be layered on top of operations without fundamentally reshaping them.

Ownership made sense. Stewardship made sense. Committees made sense.

Data governance emerged to answer reasonable questions: Who owns this data? Who is responsible for its quality? Who approves its use? How do we ensure compliance with laws, regulations, and internal policies? The answers were encoded into operating models built around domains, functions, and hierarchies.

Those answers were never wrong. They were just incomplete.

Legacy governance assumed that data could be controlled by controlling people. Assign the right roles. Define the right policies. Document the right standards. Review exceptions through the right committees. If everyone followed the process, the system would behave.

That assumption held as long as systems behaved predictably.

Over time, governance matured. Organizations invested in catalogs to document assets, glossaries to align definitions, lineage tools to trace movement, and quality frameworks to monitor correctness. Capability models emerged to professionalize the discipline and establish common language across industries. Governance became a practice – sometimes bureaucratic, sometimes effective, often uneven, but widely recognized as necessary.

Yet even at its most mature, legacy governance was fundamentally **retrospective**.

It observed data after it was created. It reviewed usage after access was granted. It audited decisions after outcomes occurred. It relied on periodic checkpoints rather than continuous enforcement. Governance was something you did *to* data, not something embedded *within* how data moved and transformed.

This was not a flaw or accidental. It was a design choice shaped by technological and staffing limits.

Data pipelines were slow. Integration was brittle. Metadata was incomplete. Enforcement at runtime was difficult. Human review was the only viable control mechanism, and humans could not be everywhere at once. Governance optimized for what was possible, not for what was ideal.

Then the environment changed.

Data stopped behaving like an asset you could inventory and started behaving like a flow you could only influence. Models began generating outputs that were not directly traceable to a single source. Automation entered decision paths previously assumed to be human-only. Context became as important as content. And suddenly, governance models built on static assignment and episodic review found themselves misaligned with reality.

The failure was subtle at first.

Governance artifacts multiplied, but their influence did not. Policies existed, but exceptions became routine. Stewardship roles were assigned, but authority remained unclear. Committees reviewed issues, but only after momentum had already carried decisions forward. Governance became a parallel system – adjacent to operations, respected in theory, bypassed in practice.

This is where many organizations still operate.

They have governance programs that are well-intentioned, staffed by capable professionals, and supported by tools that document and report. Yet those programs struggle to answer the questions leaders now ask: Who approved this AI-driven decision? Why did the model behave this way? Who is accountable when outcomes cross ethical, legal, or reputational lines?

The honest answer is uncomfortable: the governance model was never designed to answer those questions in real time.

It was designed to establish control through organizational structure rather than through the continuous evaluation of actions as they occurred.

It assumed that organizational roles were stable, responsibilities changed slowly, and decisions flowed through predefined processes that humans could observe, review, and approve. It did not anticipate a world in which a single person – or a single AI agent – could consume information, transform it, invoke other systems, make recommendations, authorize actions, and execute transactions within seconds.

Most importantly, it never anticipated that governance itself would have to operate at the speed of execution rather than at the speed of oversight.

This does not mean governance failed. It means governance succeeded at exactly what it was designed to do.

And nothing more.

The mistake organizations make today is trying to stretch these models beyond their design limits. They add more committees. They write more policies. They purchase more governance tools. They reorganize reporting structures and rename stewardship roles, hoping that incremental refinement will somehow produce a step change in effectiveness.

It won't.

The problem is not that organizations lack governance.

The problem is that governance is still positioned beside operations rather than inside them.

That distinction is subtle, but it changes everything.

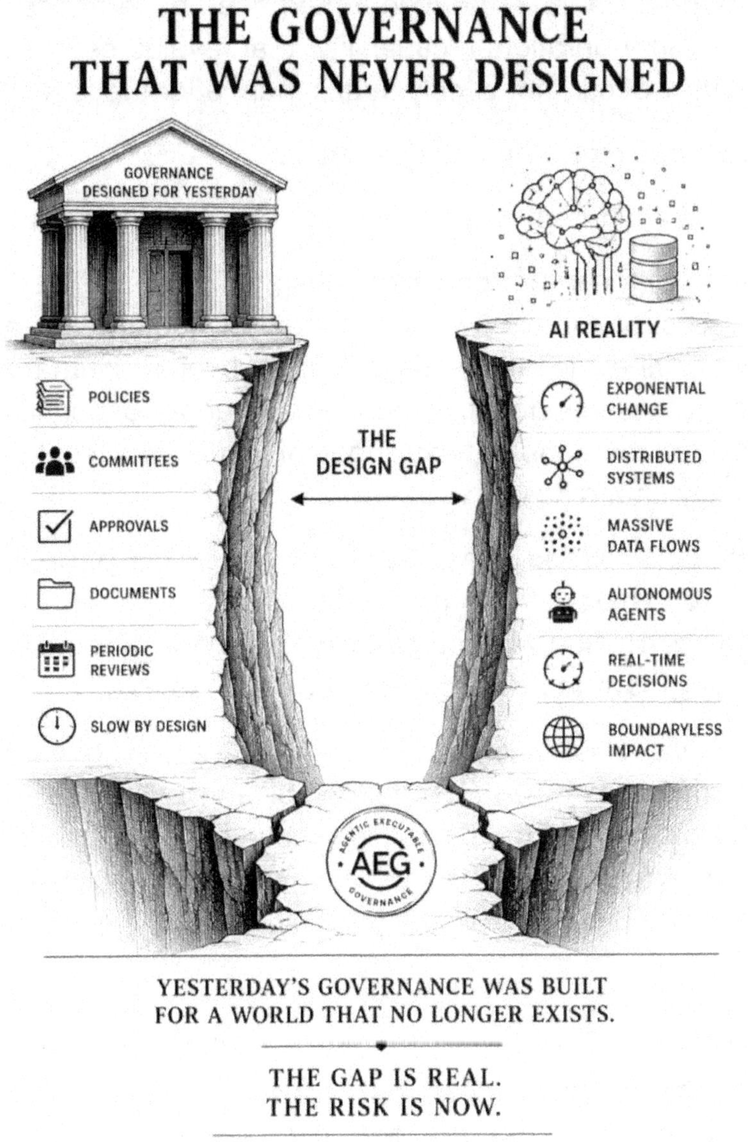


*Figure 1-4-1*

*: The Governance that was never designed*

For decades, governance has been treated as an oversight function – a mechanism that reviews, approves, documents, and reports. In an AI-driven enterprise, governance must become an operational capability that evaluates decisions before they occur, not simply audits them after the fact.

This requires more than modernization.

It requires a different way of thinking about accountability itself.

Accountability can no longer be assigned once and assumed forever because the activities that create value and risk are no longer confined to stable organizational roles. Humans and intelligent systems now collaborate, exchange responsibilities, and make decisions in ways that are fluid, contextual, and constantly changing.

> **"For decades, organizations governed people and trusted their judgment. In the age of AI, organizations must govern decisions because judgment is no longer confined to people."**

Governance must become just as adaptive. The governance we inherited was never intended to solve this problem. That is not an indictment of the past. It is simply recognition that every generation eventually reaches a point where its institutions must evolve to match the realities they govern.

Artificial intelligence has brought us to that point.

The institutional contract that governs data, decisions, and accountability must now be rewritten – not because governance no longer matters, but because it has never mattered more.

The chapters that follow introduce a different way of thinking about governance. They begin with a simple premise:

The future of governance will not be defined by who owns the data.

It will be defined by who is responsible for every action taken with it.

That begins with a new social contract.

Notes:

*Figure 1-5-1: A New Social Contract for Data*

# Chapter 5: A New Social Contract for Data and Decision Making

Every system that endures eventually renegotiates its social contract.

Not in a single moment, and rarely by unanimous agreement – but through pressure, failure, and the slow realization that old rules no longer produce fair or functional outcomes. These renegotiations are not philosophical exercises. They are practical necessities, driven by the gap between how a system actually behaves and how it is supposed to behave.

Data and its use have reached that point where a new social contract for data must be established.

For decades, organizations operated under an implicit contract about information. Data would be created by one group, maintained by another, and consumed by many. Responsibility could be assigned in advance. Oversight could be centralized. Accountability could be documented and enforced through roles that remained stable over time.

That contract assumed that information followed a predictable lifecycle. Data was created, stored, reported, and eventually archived. The participants were known, responsibilities were assigned in advance, and governance relied on relatively stable organizational boundaries.

That world no longer exists.

The assumptions behind governance changed gradually. The assumptions behind data changed overnight. The technology evolved faster than the governance model designed to contain it.

The same individual may discover data in the morning, transform it before lunch, enrich it with AI in the afternoon, approve its publication before the day ends, and trigger downstream decisions without ever realizing they have crossed multiple governance boundaries.

Legacy governance treats this as a problem of discipline.

It assumes that if people were better trained, more compliant, or more closely monitored, the system would behave. But this framing misses the deeper issue. The problem is not that people are acting irresponsibly. It is that the system assigns responsibility based on identity rather than activity.

**We ask who someone *is*.**

**Not what they are *doing*.**

For generations, this worked because the most important decisions were still being made by people. Organizations documented processes, defined roles, and established approval paths, but much of the actual decision-making occurred within the experience, judgment, and intuition of individual operators. Those mental models were rarely written down, and they rarely needed to be.

Organizations hired trusted professionals because they expected them to exercise judgment where documented procedures inevitably ended.

That assumption shaped governance for decades.

Artificial intelligence changes that equation.

When decision-making shifts from the human mind into software, undocumented judgment can no longer be assumed. The reasoning that once lived inside a person's experience must now be made explicit. Policies, context, authority, and acceptable behavior can no longer remain implicit organizational knowledge. They must become operational knowledge – explicit, machine-readable, and continuously evaluated before decisions are executed.

Job titles, organizational charts, and committee memberships have become convenient proxies for accountability. Yet accountability has never truly belonged to titles. It belongs to the activity performed at a particular moment.

A person retrieving information is not exercising the same authority as a person modifying it. A person modifying data is not performing the same function as one approving its release. A person approving its release is not making the same decision as an AI agent executing that approval automatically. Treating all of those activities as though they belong to a single organizational role obscures risk instead of controlling it.

This is where the old social contract breaks down.

Modern data environments require governance that recognizes activity states, not static positions. Responsibility must attach to activities rather than titles. Oversight must respond to behavior rather than hierarchy. Accountability must be evaluated continuously rather than assumed permanently.

This is not a call to eliminate structure. It is a call to redesign it.

A functional social contract for data acknowledges that everyone interacts with information differently depending on what they are doing. It recognizes that governance cannot remain centralized in a committee while operations become decentralized across humans and AI agents.

Under this new contract, governance becomes less about permission and more about posture.

- It shifts from asking, **Who is allowed?**
- to asking, **What is happening?**

From assigning ownership once... to evaluating responsibility continuously.

From documenting intent... to governing execution.

This reframing does not weaken accountability. It strengthens it.

When responsibility is tied to activity, decisions become auditable in context. When oversight is embedded in workflows, governance moves at operational speed. When roles are defined by function rather than title, organizations gain flexibility without losing control.

The old contract promised certainty.
The new contract offers resilience.

A functioning social contract for data rests on three structural conditions:

1. Shared Purpose – data exists to create benefit
2. Visible Boundaries – acceptable use is explicit and understandable
3. Continuous accountability – responsibility follows activity in real time.

Without these conditions, trust becomes symbolic rather than operational. A contract is only meaningful if it can be enforced. In an AI-driven enterprise, enforcement cannot depend upon periodic review. It must exist within execution itself. This requires more than redefining governance responsibilities. It requires redefining the operating roles through which governance is exercised.

*Figure 1-5-2: The Chain of Operational Trust*

The Guardian, Maker, and Seeker framework provides that operational layer, translating governance principles into repeatable actions that humans and intelligent systems can execute consistently.

**Instead of describing governance, it enables governance.**

In a world where data and AI operate continuously, resilience matters more than rigidity. The ability to adapt governance to context – to recognize when risk changes, when authority shifts, when human judgment must intervene – is what separates organizations that scale responsibly from those that lurch from crisis to crisis. And it is what will distinguish the trusted enterprise from the merely automated one.

This chapter marks the turning point.

The fracture has been diagnosed. The limits of inherited governance have been named. What comes next is not a rejection of responsibility, but its redesign, one that accepts the realities of modern data systems and builds accountability into their very operation.

- Legacy governance asks: **Who are you?**
- Transitional governance asks: **What are you doing?**
- Executable governance asks: **Who are you, what are you doing, and under what circumstances?**

The chapters that follow make this new contract explicit. They redefine governance around activity rather than hierarchy. They introduce operational roles designed for both humans and AI agents. They demonstrate how accountability can follow every governed transaction rather than every organizational title. Most importantly, they show how governance can become something it has never been before: An operational capability that executes at the same speed as the decisions it governs.

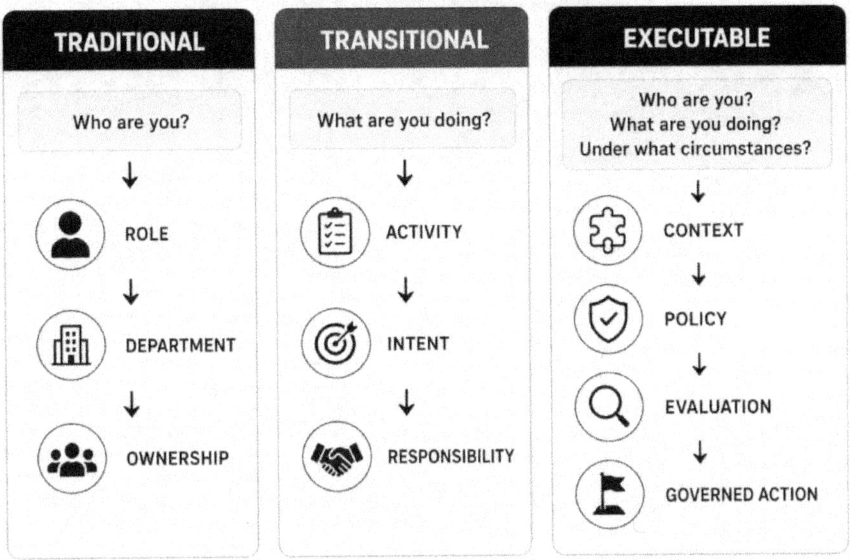


*Figure1-5-3: Role-based Governance*

For centuries, governance has been organized around who people were. The future will organize governance around what they are doing. That change sounds deceptively small. It may become one of the most significant shifts in the history of information management. The remaining chapters explore how that shift can be designed, implemented, and ultimately trusted.

# Part II: Accountability States for the Age of AI

## Standing on Strong Foundations

Legacy governance didn't fail.

It accomplished exactly what organizations needed during the Information Age. Data governance councils, executive committees, data owners, data stewards, security professionals, architects, compliance officers, privacy teams, and operational leaders created the discipline that allowed organizations to trust their information and manage risk at enterprise scale.

The enterprise we know today could not exist without those contributions.

## Responsibility follows activity.

Legacy governance assumed that responsibility could be attached permanently to organizational roles. Data owners owned. Data stewards stewarded. Consumers consumed. Governance became an extension of the organizational chart because the work itself was relatively stable.

That assumption simply won't stand up to today's reality.

The same person may retrieve information in one moment, transform it in the next, approve its publication moments later, and rely on an AI agent to execute the final transaction. Each of those activities carries a different level of authority, different risks, and different accountability.

Yet legacy governance often treats all these activities as though they belong to a single role simply because they are performed by the same individual.

This is where governance loses precision.

Responsibility does not belong to a person indefinitely.

**Responsibility belongs to the activity being performed.**

That principle becomes the foundation of the Guardian–Maker–Seeker (GMS) Framework.

Every interaction with information ultimately falls into one of three fundamental accountability states.

- **Information is protected.**
- **Information is changed.**
- **Information is used.**

From those three activities emerge three operational governance states:

- **Guardian** – Protects information, policies, and organizational trust.
- **Maker** – Creates, transforms, and maintains information.
- **Seeker** – Discovers, accesses, and applies information to create value.

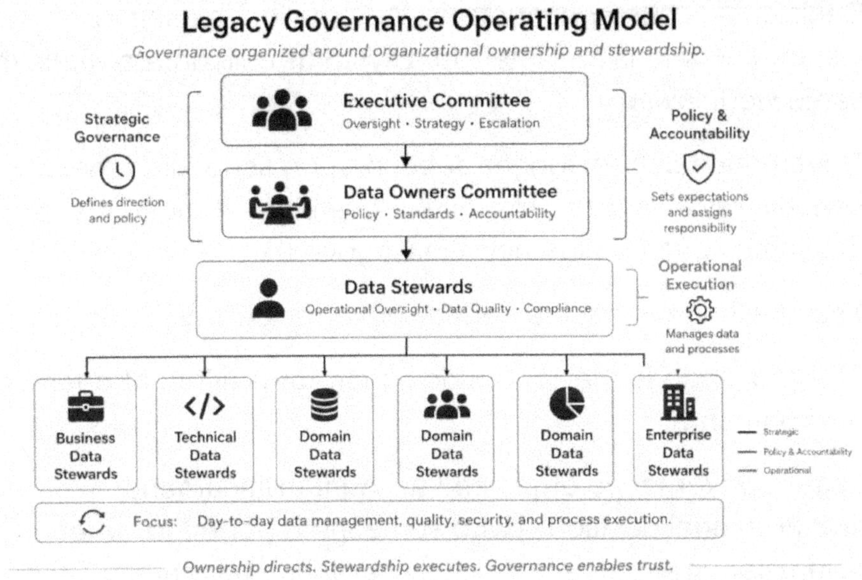


*Figure 2-0-1: Legacy Governance Operating Model*

These are not job titles. They are not departments. They are not committee assignments. They are operational accountability states through which every governed transaction passes. Once governance is organized around activity rather than identity, accountability becomes contextual, policies become executable, and trust can be established continuously rather than assumed periodically.

This is the transition from legacy governance to Agentic Executable Governance.

## Accountability States for the Age of AI

Artificial intelligence changes the operating environment fundamentally.

Decisions occur continuously. Automation executes actions instantly. Humans and intelligent systems collaborate within the same workflows.

Governance can no longer depend upon scheduled meetings, periodic reviews, or retrospective oversight. It must operate at the same speed as the activities it governs.

The challenge is not that legacy governance was incorrect.

The challenge is that it was designed for a different operating environment.

Legacy governance organized accountability around organizational position – who someone reported to, what committee they served on, and where they appeared within the enterprise hierarchy.

Agentic governance organizes accountability around activity.

It asks a different question.

Not: **Who is this person?**

But: **What are they doing right now?**

That distinction changes everything.

As people and AI agents move through the lifecycle of information, they naturally transition between protecting it, changing it, and using it. Governance must recognize those transitions and adapt in real time.

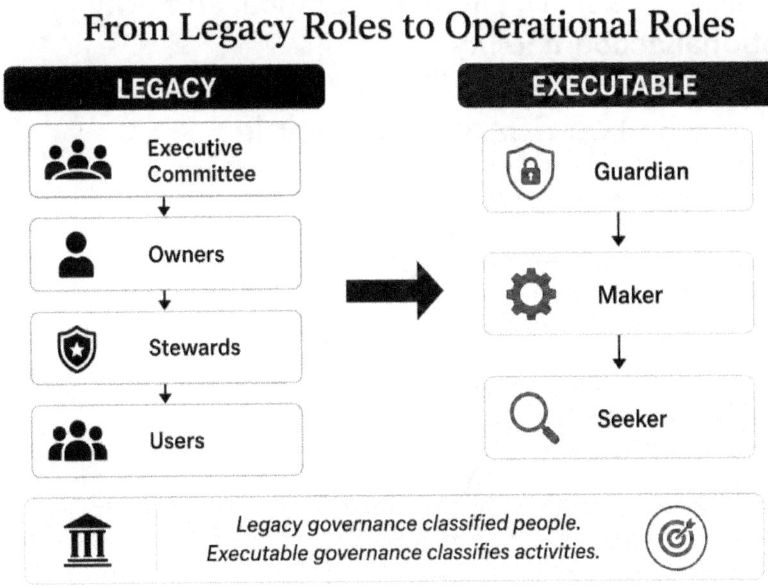


*Figure 2-0-2: From Titles to Activities*

The Guardian–Maker–Seeker Framework provides that operational model.

Rather than assigning responsibility permanently to organizational titles, it assigns accountability dynamically to the activity being performed. Governance becomes embedded within execution rather than applied after the fact.

The chapters that follow examine each accountability state individually before bringing them together as a unified operating model for the age of AI.

Governance is no longer shifting from hierarchy to automation. It is shifting from **hierarchical authority** to **situational accountability**.

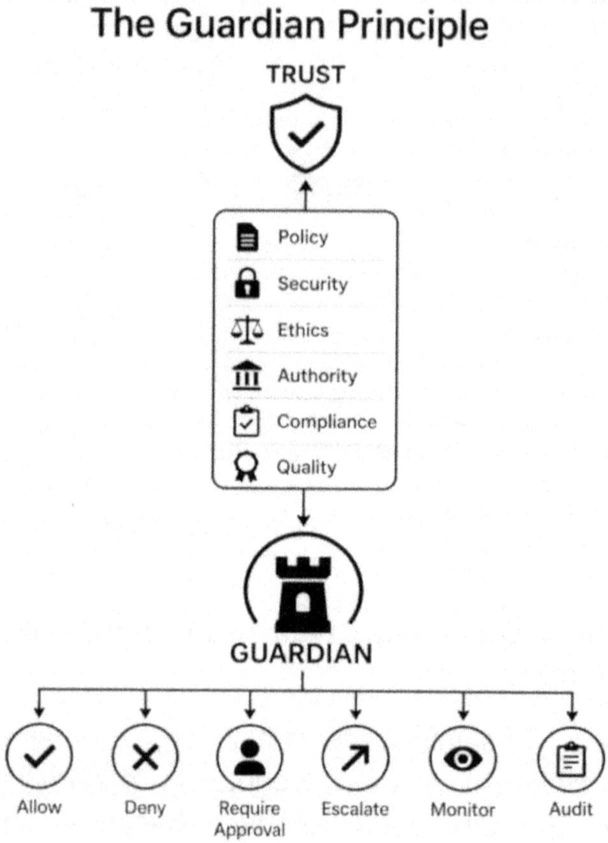


*Figure 2-6-1: The Guardian Principle*

# Chapter 6: The Guardian Principle

## Section 1: Every civilization creates Guardians.

Not because it distrusts people. Because it values what must be protected.

- Parents guard children.
- Judges guard justice.
- Librarians guard knowledge.
- Cybersecurity teams guard infrastructure.
- Boards guard shareholder interests.
- Governments guard sovereignty.
- Doctors guard patient safety.
- The military guards national security.

Every enduring institution creates Guardians. The enterprise is no different.

### The Guardian Existed Before Technology

This is an important point.

Guardians existed before computers. They existed before databases. They existed before AI.

**Technology changes what is protected.**

**It does not change why protection exists.**

Guardians are a universal governance concept. But what do Guardians protect today?

· Trust · Meaning · Policy · Authority · Ethics · Security · Integrity · Compliance · Organizational Intent

Data is simply one manifestation.

> **"The Guardian does not exist to stop progress. The Guardian exists to ensure progress remains worthy of trust."**

All civilizations that have endured have created Guardians – Militaries, Priests, Historians, Governments.

## Section 2: The Guardian Defines the Boundary

Every system that operates at scale eventually confronts the same question: who decides what must not happen and who decides what should be allowed?

Not who optimizes. Not who builds. Not who benefits. But who draws the boundary - and holds it, even when doing so is inconvenient, unpopular, or expensive.

We used to assign these decisions to the role of "Data Owner." But in modern data and AI environments, that responsibility belongs to the Guardian.

The Guardian is not a job title. It is not a committee. It is not a department. It is a **state of** accountability that emerges when oversight is required. A person may act as a Guardian in one moment and not in the next. A system may enforce Guardian constraints continuously, even when no human is actively watching. The Guardian is not a reviewer of decisions. The Guardian is the role through which constraints are defined so that human and AI systems can act without violating trust.

The Guardian defines the boundaries inside which value can safely be created.

*Figure 2-6-2: What the Guardian Protects*

Legacy governance models treated guardianship as a static assignment. Data owners, stewards, compliance officers, and risk committees were designated in advance and expected to provide ongoing oversight across broad domains. Authority was attached to position, and responsibility was assumed to persist regardless of context.

That approach collapses under AI.

In environments where data flows continuously and AI models participate in decision-making, **Oversight** cannot be episodic.

But it must be **Situational**. The Guardian role activates when risk is present - when data is accessed beyond its intended context, when transformations alter meaning, when automated decisions carry ethical, legal, or reputational consequences.

## Section 3: Constraint Creates Trust

Guardianship is often misunderstood as control.

It is not.

Guardianship is constraint.

Constraint is what allows freedom to exist safely.

Without boundaries there is no trust. Without trust there is no sustainable innovation.

A Guardian does not decide what the organization should do. A Guardian defines what the organization must never do - and under what conditions intervention is required. This includes setting boundaries around privacy, security, fairness, explainability, regulatory compliance, and acceptable use. It also includes enforcing those boundaries in real time, not merely documenting them.

This is where legacy governance struggles most.

Oversight is often separated from execution.

Data Owners and Stewards review outcomes after the fact.

Risk assessments are conducted during project initiation, not continuously during operation. Ethical considerations are articulated in principle statements, then left unenforced at runtime.

When issues surface, they are escalated through committees that lack the authority or context to act decisively.

The Guardian role redefines this relationship.

Governance becomes embedded within the workflow itself.

*Figure 2-6-3: The Guardian Trust Chain*

## Section 4: From Oversight to Embedded Governance

In a modern governance model, Guardianship is embedded in workflows. Constraints are encoded into systems. Decisions are evaluated against policy at the moment they occur. Human Guardians intervene when thresholds are crossed, ambiguity arises, or judgment is required beyond what automation can provide.

The Guardian establishes far more than policy. Guardians establish the entire chain through which organizational trust is earned.

This does not eliminate human oversight. It makes it effective.

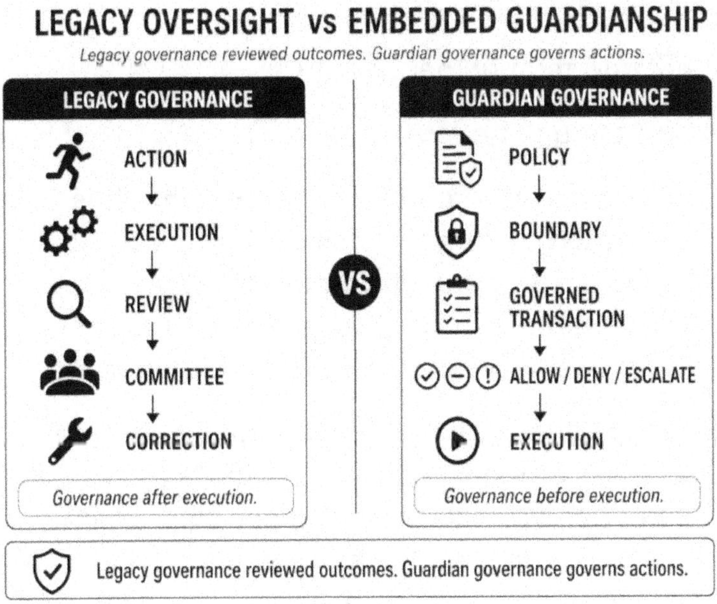


*Figure 2-6-4: Legacy Oversight vs Embedded Guardianship*

## Section 5: Human Guardians and Machine Guardians

A Guardian may be

- a privacy officer approving sensitive data use,
- a legal advisor defining regulatory boundaries,
- a security architect setting access constraints,
- a business leader accountable for ethical outcomes,
- or another accountable authority.

What unites these actors is not their title, but the function they perform in that moment - protecting trust by limiting action.

Their title is secondary.

Their accountability state defines them.

Trust is not built through permission.

Trust is built through performance.

**"Trust is not created by permission. It is created through consistent governed execution. Organization earn trust because every action is evaluated, every boundary is enforced, ad every exception is accountable."**

Organizations often believe that trust comes from restriction - fewer users, fewer tools, fewer paths to data. In reality, trust comes from consistent enforcement of clearly defined boundaries. When stakeholders understand what is allowed, what is prohibited, and how violations are handled, confidence increases even as access expands.

Guardians make this possible.

They provide the assurance that innovation can proceed without eroding values or exposing unacceptable risk. They enable speed by removing ambiguity. They protect the organization not by slowing it down, but by making its limits explicit and enforceable.

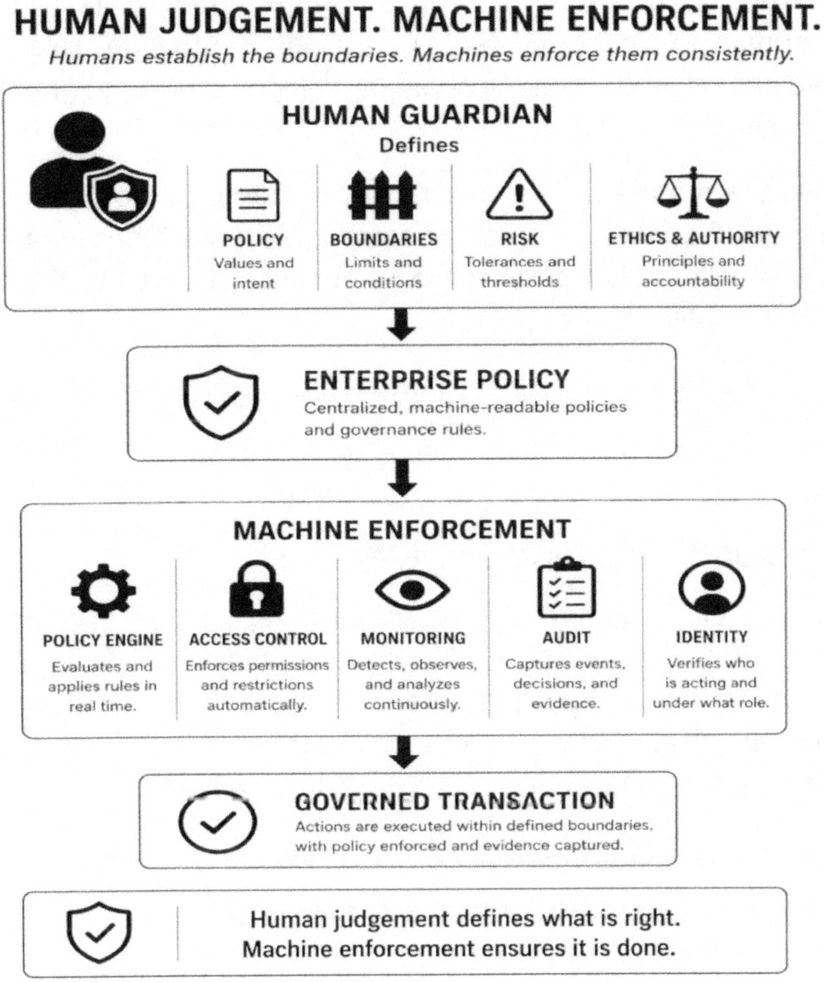


*Figure 2-6-5: Human Judgment. Machine Enforcement*

Importantly, the *function* of Guardianship is not always performed manually and is not always human.

In AI-enabled environments, many Guardian functions are automated. Access controls, policy engines, monitoring systems, and audit mechanisms operate continuously, flagging or blocking actions that violate defined constraints. These systems act as persistent Guardians, enforcing rules at machine speed while escalating edge cases to human judgment.

They are not the Guardian. They operationalize the Guardian. Human Guardians remain accountable. Machines enforce consistently. This partnership is essential for scale. But the human is still the accountable party.

Together they create governance capable of operating at the speed of AI.

## Section 6: Holding the Line

Human Guardians excel at interpreting context, weighing tradeoffs, and adapting to novel situations. Automated Guardians excel at consistency, scale, and immediacy. Together, they form a governance layer that can operate at the same speed as data and AI systems themselves.

Without this layer, organizations default to panic.

They restrict access broadly rather than govern precisely. They pause initiatives rather than manage risk dynamically. They react to failures rather than preventing them. Guardianship replaces this reactive posture with intentional design.

**The Guardian role is the foundation of the new social contract for data.**

In public discussions, stewardship of data is often described in ethical terms — fairness, transparency, or public trust. These principles are essential, but they do not explain how a system behaves when confronted with real decisions.

The Guardian exists to answer that question. The Guardian is the institutional actor responsible for translating values into boundaries:

- Policies become constraints.
- Standards become thresholds.
- Risk tolerance becomes intervention logic.

The social contract becomes real not when it is written, but when it is enforced.

It ensures that as roles become fluid and systems become autonomous, responsibility does not dissolve. Someone is always accountable for holding the line. Increasingly, machines help enforce those boundaries with consistency and speed, but accountability ultimately remains a human responsibility. The Guardian establishes the conditions under which governance decisions are made. In modern enterprises, those conditions increasingly become machine-readable policies that can be enforced automatically. The Guardian remains accountable for the boundary. The systems simply ensure that boundary is applied consistently at the speed of execution.

The next chapters will introduce the other accountability states that complete the model. But none of them function without the Guardian.

Before anything can be built…

Before any information can be transformed…

Before any intelligence can act…

Someone, or Something operating under accountable human authority, must decide whether that action remains worthy of trust.

That is the purpose of the Guardian.

Everything else depends upon it.

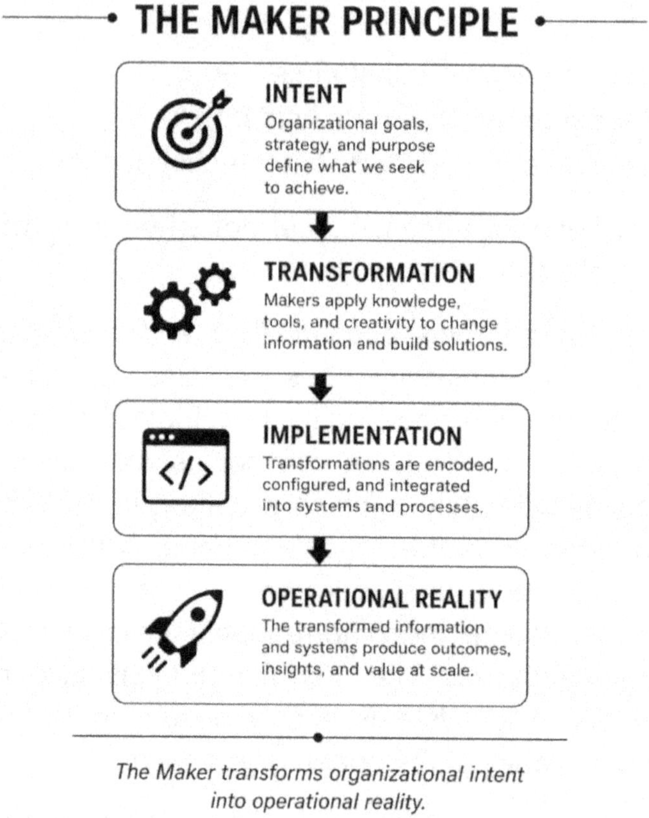


*Figure 2-7-1: The Maker Principle*

# Chapter 7: The Maker Principle

## Section 1: The Maker Principle

If the Guardian defines the boundary, the Maker walks it.

Not recklessly. Not maliciously. But deliberately, as part of the work of creating value. Makers are the people – and increasingly the intelligent systems – that transform information into something new. They transform data, train models, modify logic, and reshape information so it can serve new purposes. These activities may introduce new risks, alter existing boundaries, or require Guardian review before they proceed.

Every organization depends on Makers.

They build pipelines, write code, configure systems, engineer features, clean data, enrich context, fine-tune models, and automate decisions. Without Makers, nothing evolves. No insight is produced. No innovation occurs.

The Maker is the engine of progress.

And the source of most governance failure.

Legacy governance models treated Makers as implementers. They were expected to follow standards defined elsewhere, operate within controls set by others, and execute decisions that had already been approved. Responsibility for outcomes was assumed to rest upstream - with owners, stewards, or committees.

That separation no longer holds up.

In modern data and AI environments, the act of making *is* the decision. Transformations encode assumptions. Feature selection embeds bias. Model tuning prioritizes outcomes. Automation hardens judgment into logic. Once deployed, these choices operate at scale, often invisibly, and often without further human review.

> **"Makers do not merely execute intent.**
> **They instantiate it."**

This is why the Maker role must be treated as a state of accountability, not a technical function.

A person may act as a Maker when writing a SQL query that permanently alters a dataset, configuring a pipeline that feeds a model, adjusting thresholds that change how decisions are made, or deploying an AI model into production. In each case, they are not just building - they are shaping meaning.

Yet governance has historically struggled to engage Makers effectively.

Controls were designed around access rather than action. Reviews occurred before or after changes, not during the process. Standards were documented but not enforced at the moment of transformation. The result was a gap between governance intent and operational reality.

## Section 2: Why Legacy Governance Failed Makers

Makers learned to move fast around governance, not through it.

This was not defiance. It was necessity. When governance introduced friction without clarity, progress depended on bypassing it. Over time, this dynamic hardened into culture. Governance became something to avoid. Makers became accustomed to carrying responsibility without authority, accountability without support.

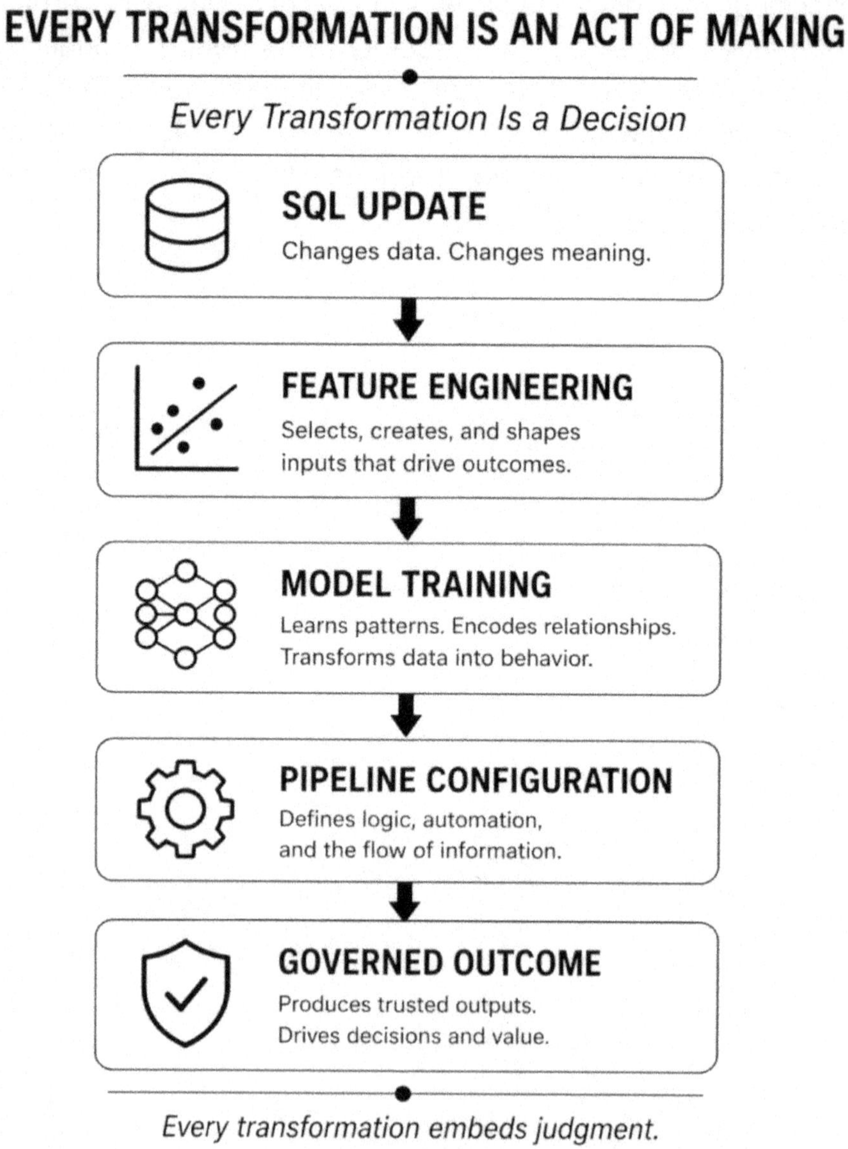


Figure 2-7-2 Every transaction is an act of Making

AI intensifies this tension.

Models amplify the impact of Maker decisions. A single configuration change can affect thousands of downstream outcomes. A subtle shift in training data can alter behavior in ways that are difficult to detect and harder to explain. Once deployed, models act continuously, executing the logic Makers have embedded long after the moment of creation has passed.

In this environment, treating Makers as neutral implementers is dangerous.

The Maker role requires explicit recognition, protection, and constraint. Makers need clear boundaries within which they are empowered to operate - and mechanisms that surface risk at the moment it is introduced, not months later during an audit.

This is where the Guardian and the Maker intersect.

**Effective governance does not slow Makers down**. Guardian constraints must be visible at the point of making. Policies must be enforceable during transformation. Feedback must be immediate. When a Maker crosses a boundary, the system should respond in context - not through escalation after the fact.

Guardians must also be prepared to respond when Makers encounter legitimate boundary questions. Effective governance depends not only on clear constraints, but on timely decisions. Innovation slows when review cannot keep pace with change. This partnership changes everything.

Makers gain confidence because expectations are clear. Guardians gain leverage because controls are embedded. The organization gains speed because ambiguity is reduced. Accountability becomes traceable because actions are logged, evaluated, and understood in context.

## Section 3: Makers Are Not All Engineers

Importantly, not all Makers are technical. Every time someone changes meaning, they become a Maker.

A business analyst modifying logic in a reporting tool, a product manager configuring decision rules, a marketer adjusting targeting criteria - each is acting as a Maker when they change how data is interpreted or used. AI expands the Maker role beyond engineering, making governance everyone's concern without making everyone a gatekeeper.

The greatest risk is not that Makers will act irresponsibly. It is that they will act without visibility.

When Maker activity is opaque, governance defaults to restriction. Access is limited broadly rather than managed precisely. Innovation slows. Frustration rises. Shadow systems proliferate.

When Maker activity is governed in context, the opposite occurs.

Constraints become guides rather than obstacles. Oversight becomes collaborative rather than punitive. Makers are trusted because their actions are accountable, not because they are assumed to be careful.

The future is built by Makers. Every model. Every pipeline. Every workflow. Every autonomous decision. Every governed transaction. Every data product, every AI-enabled workflow, every automated decision begins here.

Makers create the future one transformation at a time.

Every one of these actions changes the enterprise. That is why the act of Making is never a purely technical activity. It is an act of governance.

The important insight is that the act of Making is fundamentally one of changing the organization. And the role is not generally recognized or supported through operations or governance. The question is not whether Makers should exist - they already do. The question is whether organizations are willing to acknowledge the transformational power of *making* and govern it accordingly.

Every transformation is a decision. Transformation changes meaning. Meaning is the information that runs the business. Understanding the role of Maker, and its pervasiveness through your organization, is crucial to provide the right training, support and appropriate authority.

The next chapter introduces the final role in this system - the Seeker. The consumer of data and AI outputs, whose questions drive demand and whose trust ultimately determines whether the system succeeds.

But without Makers, there is nothing to seek.

# Chapter 8: The Seeker

Every system is ultimately judged not by what it can produce, but by what people choose to trust.

That judgment belongs to the Seeker.

Seekers consume information and AI-generated outputs. They ask questions, request insights, review recommendations, and act on what systems return to them. Executives scanning dashboards. Clinicians relying on decision support. Analysts exploring trends. Frontline employees following automated prompts. Customers receiving personalized offers.

Without Seekers, data has no purpose and AI has no impact.

Yet governance has historically treated the Seeker as passive – the final stop in a linear flow. Data is created, transformed, approved, and then delivered to the user. Responsibility was assumed to diminish as information moved downstream. If something went wrong, attention focused upstream – on data quality, engineering, or governance controls.

That assumption no longer holds.

In modern data and AI environments, seeking is not passive. The questions Seekers ask shape outcomes. Queries define scope. Prompts influence behavior. Interpretation determines action. Makers introduce both value and risk by changing systems. Seekers determine whether those changes become trusted decisions.

Seeking is an act of agency.

The moment information is used, it is no longer abstract. Context becomes real. Meaning is applied. A decision path begins. Whether the organization acknowledges it or not, the Seeker exercises authority at the point of use.

This is where governance often fails quietly.

Organizations focus on controlling access rather than guiding use. They provide powerful data and AI tools with minimal context, assuming users will interpret outputs correctly. They rely on training, disclaimers, and policy documents rather than embedding guidance at the moment decisions are made.

When misuse occurs, it is framed as user error rather than as a system design failure.

But Seekers operate under pressure.

They are expected to move quickly, make decisions, and deliver results. They trust systems because those systems have been approved, deployed, and institutionalized. Few Seekers have the time or expertise to interrogate lineage, evaluate model limitations, or question assumptions embedded upstream.

They rely on the signals the system provides – confidence, explanation, fluency, and organizational endorsement.

When those signals are misleading, trust erodes.

For this reason, the Seeker role must be governed not through restriction, but through **contextual enablement**.

Seekers need to understand what a result represents, how it should be used, and where its limits lie. They need clarity about confidence and applicability, without being burdened by technical detail. Governance that ends at delivery leaves Seekers exposed and organizations vulnerable.

This does not mean turning every Seeker into a data expert or ethicist.

It means designing systems that carry responsibility forward into consumption. Guidance must appear at the moment of use. Boundaries must be visible when decisions are made. Feedback must be possible when results conflict with reality.

This is how the Seeker completes the governance loop.

Seekers are often the first to notice when systems fail, when outputs feel wrong, recommendations conflict with experience, or automated decisions produce unintended consequences. In well-designed environments, this feedback is captured and routed back into improvement. In poorly designed ones, it is ignored until trust collapses.

The Seeker is not the weakest link.

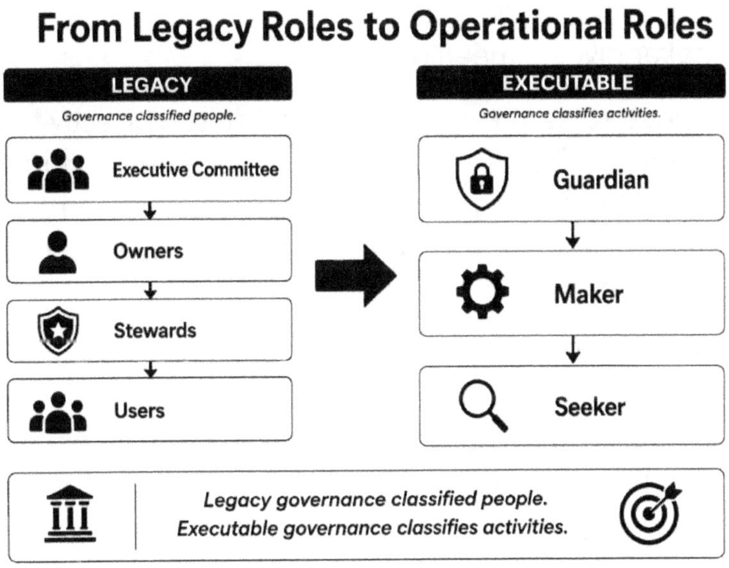


*Figure 2-8-1: From Legacy Roles to Operational Roles*

The Seeker is the legitimacy test.

AI intensifies this role. As interfaces become more natural and responses more confident, Seekers are encouraged to trust outputs more readily. Reduced friction improves usability – but may mask the lack of governance controls.

If Seekers cannot see the boundaries, they will assume none exist.

Effective governance ensures that Seekers are never asked to make blind decisions. It provides transparency without overload. It enables questioning without undermining authority. It balances usability with accountability.

Most importantly, it treats Seekers as active participants in the system, not passive recipients.

The Guardian defines the boundary.
The Maker shapes the system.
The Seeker gives it meaning.

When these roles are understood as states of activity rather than static titles, governance becomes continuous. Accountability follows action. Trust is earned through performance, not promised through policy. The social contract is honored when that trust is justified, not assumed.

These roles are not fixed stages in a process. They are accountability states that emerge whenever particular activities occur. A single person – or AI system, may move between them repeatedly within the same business process or decision. Together, the Guardian, Maker, and Seeker form a living governance system – one capable of operating at the speed of modern data and AI without sacrificing human judgment or organizational integrity.

*Figure 2-8-2: The Chain of Operational Trust*

The next part of this book turns from roles to systems – from responsibility to architecture. Because roles alone are not enough. They must be supported by foundations designed for the world we now inhabit.

# Part III – The System

Roles explain *who* is responsible.
Systems determine *whether responsibility can actually be exercised*.

Parts I and II established two essential truths. First, the governance models we inherited were never designed for the speed and behavior of modern data and AI. Second, accountability must follow activity rather than remain fixed to organizational titles or committees.

But roles alone do not change outcomes.

A Guardian without enforceable constraints is symbolic.
A Maker without contextual feedback is dangerous.
A Seeker without visibility is exposed.

If governance is to function at scale, it must be carried by systems that record context, enforce boundaries, and adapt as behavior changes. The shift to AI-capable environments is not a technology upgrade. It is an architectural pivot.

This part of the book examines that pivot.

Not as a catalog of technologies, but as a reordering of how data is organized, understood, and made usable. The systems described here are not optional enhancements. They are the structural conditions to make accountable AI possible.

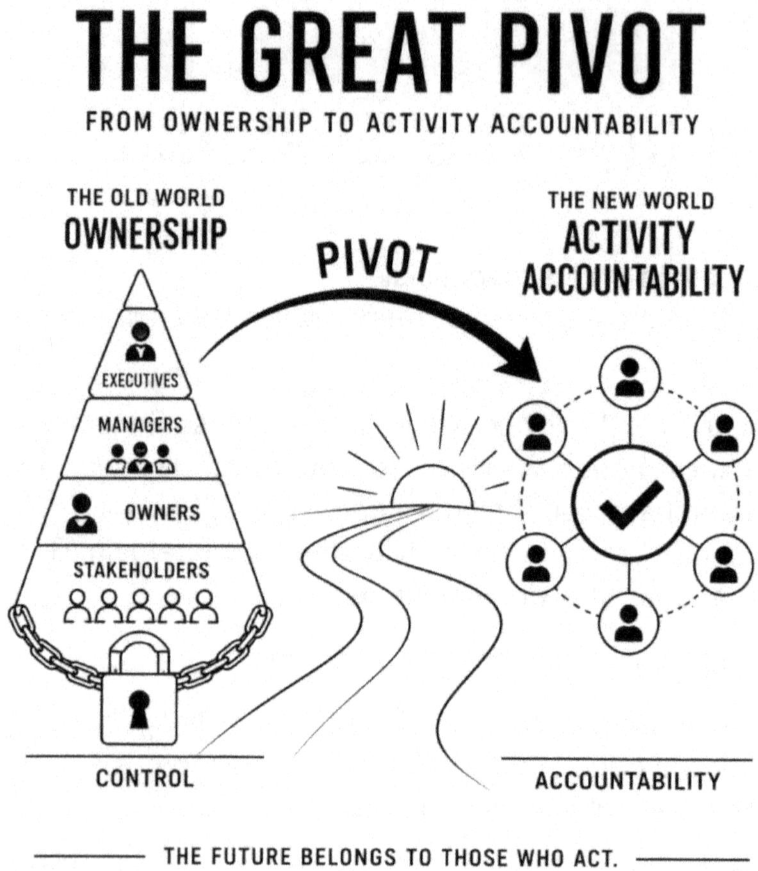


*Figure 3-9-1: The Great Pivot*

# Chapter 9: The Great Pivot

Every transformation reaches a moment when incremental improvement is no longer sufficient.

Organizations experience this moment as tension. Existing systems still function, but only with increasing effort. Controls still exist, but only through workarounds. Insights are still produced, but trust in them erodes. The cost of maintaining the old model quietly exceeds the cost of replacing it.

This is the moment of the pivot.

For decades, enterprises organized data around systems of record. Information was stored in applications designed to support transactions, reporting, and compliance. Data models reflected process flows. Governance focused on stabilizing what already existed.

This approach made sense when data changed slowly and use cases were predictable.

It fails when data becomes the raw material for AI.

AI systems do not use data the way traditional reports do. They require context, relationships, and meaning that extend beyond individual tables or applications. They draw inferences across sources, combine signals dynamically, and produce outputs that cannot be fully understood by tracing a single lineage path.

The problem is not volume. It is **shape**.

Most organizations still operate data estates designed to answer questions that were known in advance. AI thrives on questions that are discovered in the moment. That mismatch forces organizations into brittle workarounds - excessive data wrangling, shadow pipelines, duplicated logic, and one-off models that cannot be reused or governed consistently.

This is where the old model begins to crack.

The Great Pivot is the shift from **managing data as assets** to **operating data as a system of meaning**.

In the old model, value is created by collecting, cleaning, and reporting data. In the new model, value is created by understanding how data relates, how it is used, and how it changes over time. Context becomes first-class. Metadata is no longer descriptive - it is operational.

This pivot is often misunderstood as a tooling decision.

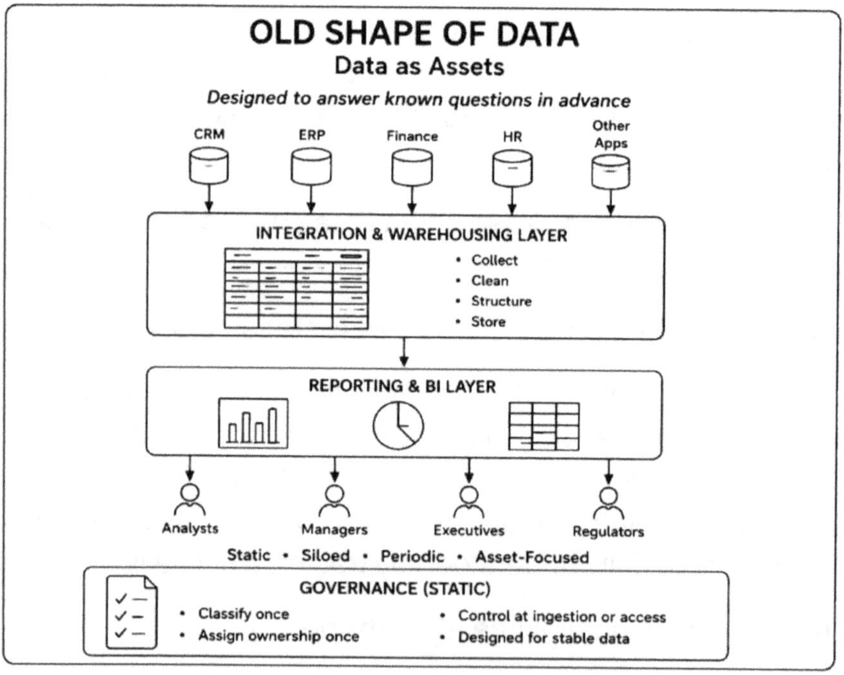


Figure 3-9-2: The Old Shape of Data

Organizations debate platforms, vendors, and architectures without first recognizing the deeper shift required. They attempt to bolt AI onto existing data warehouses, expecting new behavior from old foundations. When results disappoint, the technology is blamed rather than the structure beneath it.

But AI does not fail because models are insufficient.
It fails because the environment cannot support it.

In legacy environments, governance relies on static definitions. Data is classified once. Ownership is assigned once. Controls are applied at ingestion or access. These controls do not adapt as data is combined, repurposed, or inferred.

AI exposes this limitation immediately.

When models generate new signals, derive attributes, or infer relationships, the original governance context is lost. What was once low-risk data can become sensitive. What was once appropriate use can become harmful. Without systems that track context dynamically, governance becomes guesswork.

The pivot is unavoidable.

Organizations must move toward architectures that treat data as interconnected rather than isolated. They must capture not just what data is, but how it is used, by whom, and for what purpose. They must design environments where governance can operate continuously, not episodically.

This does not require abandoning existing systems overnight.

The Great Pivot is not a rip-and-replace strategy. It is reorientation. Systems of record remain important, but they are no longer sufficient. They feed a higher-order layer where relationships, usage, and meaning are represented explicitly.

This is the layer where AI becomes governable.

In this new model, governance is no longer something applied *to* data after the fact. It is embedded in how data is accessed, transformed, and consumed. Guardian-defined constraints accompany data throughout its use. Maker actions are evaluated in context. Seeker usage becomes visible and accountable.

The pivot changes the economics of data as well.

When context is captured and reused, data products can be shared rather than rebuilt. When meaning is explicit, discovery replaces guesswork. When trust is enforced systemically, AI initiatives move from experimentation to scale.

The alternative is familiar.

**Organizations that resist the pivot accumulate technical debt disguised as innovation**. They produce impressive demonstrations that cannot be sustained. They tighten controls reactively, slowing progress while failing to restore confidence. They remain trapped between ambition and fear.

The Great Pivot offers a different path.

It acknowledges that AI has changed the shape of data, and that governance must change with it. It accepts that accountability cannot be bolted on. It recognizes that systems, not intentions, determine outcomes. It clarifies that principles alone cannot govern machine-speed systems. A social contract for data must eventually become computable – expressed in structures that machines can interpret, enforce and audit.

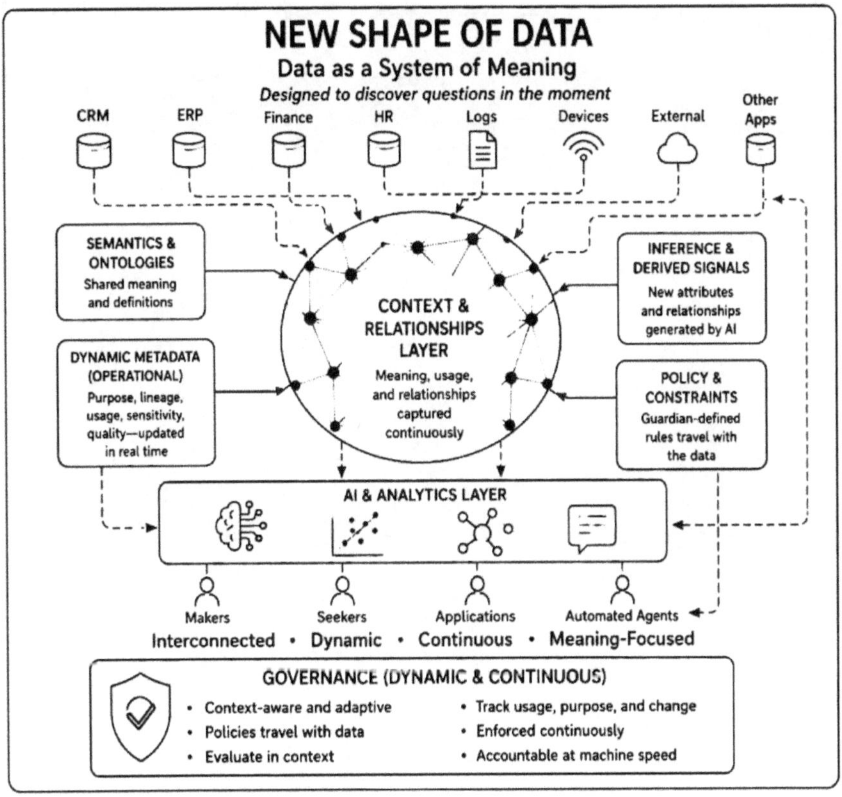


*Figure 3-9-3: The New Shape of Data*

The chapters that follow will examine the foundations of this new system - how data is mapped, how context is captured, and how governance becomes operational rather than symbolic.

The pivot has already begun.

The only remaining question is whether organizations will design it deliberately - or be forced into it by failure.

# Chapter 10: From Data Assets to Living Maps

For most of the data era, organizations treated information as property.

Data was collected, stored, cataloged, and protected. Assets were inventoried. Ownership was assigned. Value was measured by volume, coverage, and control. If data could be accounted for, it was assumed to be understood.

That assumption held as long as data remained local.

When information lived inside specific systems and supported known processes, meaning was implicit. The application supplied the context. The people using the system already understood what the data represented and how it should be interpreted. Governance focused on protecting the asset, not explaining it.

AI changed that relationship.

As data began moving freely across systems, teams, and use cases, meaning no longer traveled with it. The same data could represent different things in different contexts. What was appropriate in one setting could be risky in another. What was once clearly defined became ambiguous the moment it left its original home.

This is when data stopped behaving like an asset and started behaving like terrain.

Terrain must be navigated, not inventoried.

A list of coordinates does not tell you how a landscape fits together. It does not show elevation, proximity, or boundaries. A map does. It reveals relationships. It provides orientation. It allows people to move with purpose rather than guesswork.

Organizations now require the same shift in how they understand data.

Living maps replace static inventories. They do not merely list what exists; they show how information connects, how it is used, and how its meaning changes across situations. They evolve as the organization evolves. They reflect reality as it is practiced, not as it was documented.

This distinction matters.

Legacy data management assumed definitions were fixed. Once agreed upon, they were expected to endure. But meaning in modern enterprises is not fixed. New products emerge. Regulations change. Business models shift. Data that once carried little risk becomes sensitive. Data that once required strict control becomes broadly shareable.

Static definitions cannot keep up.

Living maps adapt. They capture shared understanding at a moment in time while remaining open to revision. They provide a common frame of reference that can be updated without rewriting every system or retraining every user.

This is the human side of the Great Pivot.

Before machines can reason responsibly, people must agree on how they understand their world. They must share a common map of the concepts, relationships, and boundaries that matter to the organization.

This is not a technical exercise – it's a governance decision.

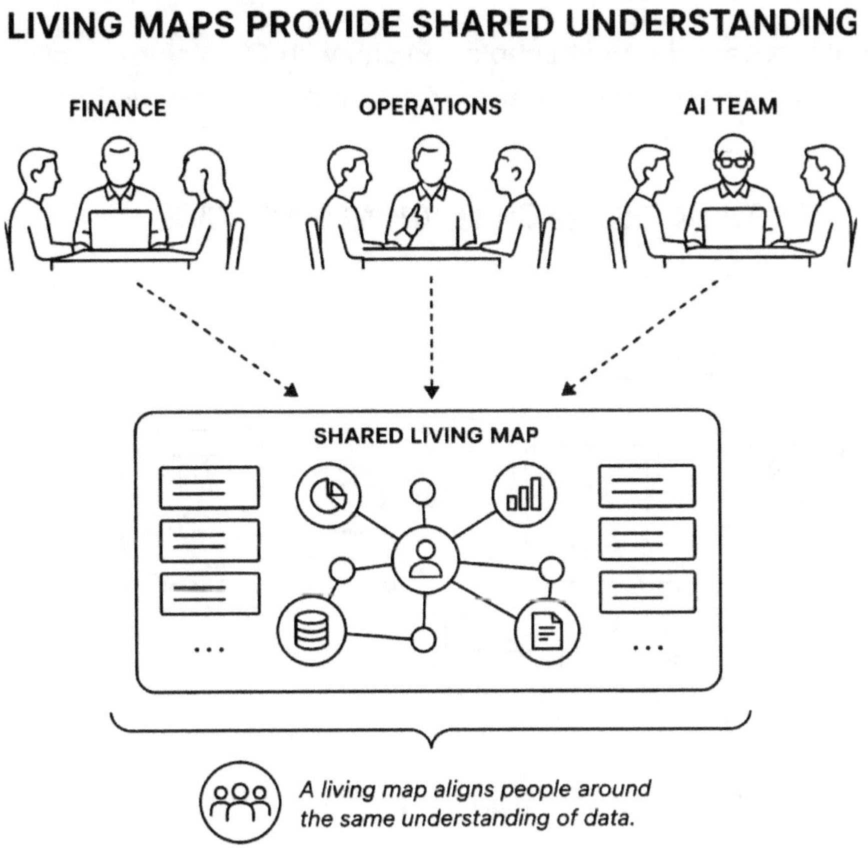


*Figure 3-10-1: Living Maps provide shared understanding*

Living maps shift governance conversations away from arguments about ownership and toward shared understanding. They replace the question "Who owns this data?" with "What does this represent here, now, for this use?"

That shift alone reduces friction, ambiguity, and risk.

But living maps are not only for people.

If governance is to operate at scale, machines must be able to use the map as well.

That is where the next chapter begins with Ontologies to align both human and machine use with common understanding.

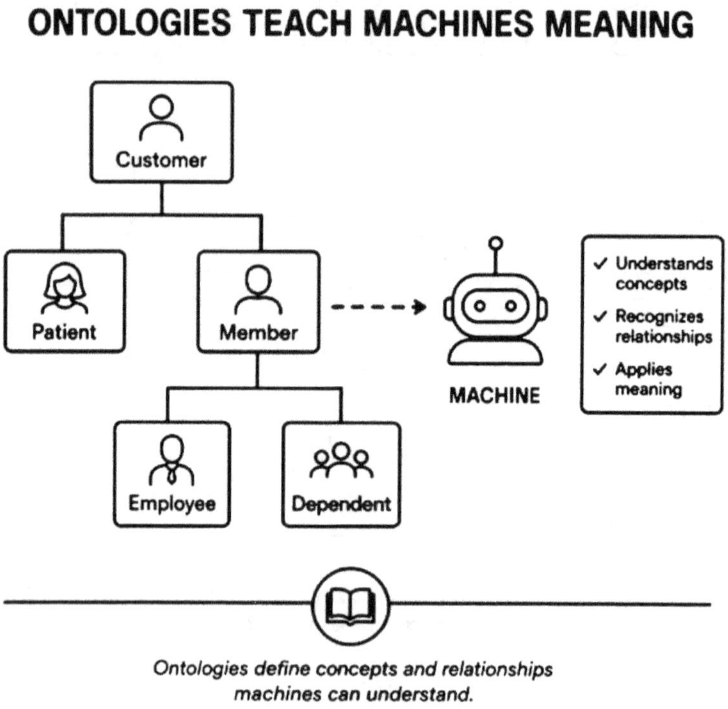


*Figure 3-11-1: Ontologies Teach Machines Meaning*

# Chapter 11: Ontologies – Teaching Machines Meaning

Once people share a map, the next challenge becomes unavoidable: machines must use it too.

AI systems do not understand meaning the way humans do. They recognize patterns, correlations, and probabilities. Without guidance, they infer context from data alone. Sometimes those inferences align with human intent. Often they do not.

This is where schemas reached their limit.

Schemas describe structure. They define fields, types, and constraints. They ensure data can be stored, moved, and processed reliably. But they do not explain what data *represents*, why it exists, or how it should be interpreted when conditions change.

Schemas assume meaning is stable. AI immediately exposes that assumption as false.

When models combine data across domains, infer new attributes, or generate outputs for new use cases, structural correctness is no longer enough. A perfectly valid schema can support a deeply flawed interpretation. Without explicit context, machines will guess - and they will guess at scale.

Ontologies exist to prevent this.

Ontologies existed long before AI. Originally a philosophical discipline concerned with the nature of being, the term is now widely understood to describe a set of concepts and categories within a subject area, together with their properties and relationships.

An ontology is a formal representation of meaning. It defines the concepts an organization cares about, the relationships between those concepts, and the rules that govern how those concepts may be interpreted together. Unlike schemas, ontologies are not tied to a specific system or dataset. They exist above the data, providing a shared semantic foundation that both people and machines can use.

Where schemas ask *how data is stored*, ontologies ask *what it means*.

In practical terms, ontologies teach machines:

- What things are, not just how they are labeled
- How concepts relate across domains
- When two records represent the same real-world entity
- How meaning changes with context
- Where boundaries and constraints apply

This distinction is subtle - and profound.

Consider a simple example. A schema may define a field called "customer." The term customer in one system may mean "someone who buys our product." In another system customer might mean "a system that pulls data via API." We may define customers as internal or external. And this definition is recorded and enforced in a rigid system schema.

An ontology defines what a customer is: a person or organization, related to accounts, consent, transactions, risk, and lifecycle. It illustrates when two records refer to the same customer, how customer roles differ by context, and how customer data may or may not be used in different situations.

Ontologies teach machines how to think about data, not just how to store it.

This is why ontologies are foundational to AI governance.

AI systems do not fail because they lack data. They fail because they lack shared meaning. Without ontologies, each model relies on the definition prebuilt into systems. It then learns its own interpretation based on patterns alone. Those interpretations drift, conflict, or violate organizational intent without warning.

Ontologies provide a counterweight to that drift.

They make context explicit. They encode assumptions that would otherwise live only in human judgment or documentation. They allow multiple systems and models to reason from the same conceptual ground, even when operating independently.

This is the principle that underpins the pivot:

**Structure organizes data.**

**Context governs its use.**

Legacy data management optimized for structure. AI requires context. Governance depends on it.

Importantly, ontologies are not static reference models.

They evolve as the organization evolves. New concepts are introduced. Relationships are refined. Rules are updated as policies, risks, and business priorities change. This adaptability is not a weakness - it is a requirement. Fixed meaning becomes a liability in dynamic environments.

Most organizations already have fragments of ontologies without realizing it.

Business glossaries, policy definitions, reference models, and data standards all attempt to capture meaning. Ontologies unify these efforts into a single, operational layer that machines can actually use.

This is the shift from documenting meaning to operationalizing it.

Ontologies do not make AI intelligent. They make intelligence accountable.

They are the bridge between human understanding and machine execution - the mechanism by which organizations teach machines how to respect meaning rather than merely process data.

The next chapter builds on this foundation.

If ontologies define meaning, knowledge graphs make meaning navigable - connecting data, metadata, usage, and decisions into a system that can be explored, governed, and trusted at scale.

*Author's note: I would be remiss if I left out this important clarification. Although I describe Ontology as being fairly flexible, a true formal ontological standard does exist. Several formats are acceptable as formal ontologies. Most of us, however, are not trained in library science and gratefully accept anything remotely ontology-like. I would rather people understand that such a thing exists and that it's useful than to struggle to stay within any rigid formula. To the ontology specialists reading this, you'll recognize that I've intentionally simplified the discipline in service of the broader argument of this book. I am aware I am not sharing the full discipline. To everyone else, I hope you can see the incredible value of ontologies and strive to utilize them.*

# Chapter 12: Knowledge Graphs – Making Meaning Navigable

Ontologies teach machines what things mean.

Knowledge graphs reveal what those meanings *do* when reality intervenes.

This distinction matters more than it first appears.

An ontology defines concepts, relationships, and rules. It establishes shared understanding. But until those definitions encounter real data, real usage, and real decisions, their implications remain theoretical. Knowledge graphs are where meaning meets consequence.

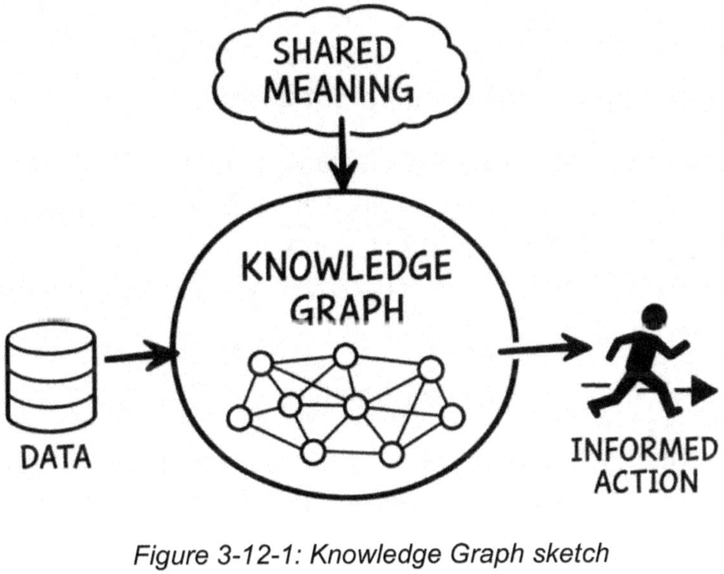


*Figure 3-12-1: Knowledge Graph sketch*

Knowledge Graphs help us visualize connections between information. They are more than diagrams. They are more than databases with nodes and edges. Knowledge graphs are the operational environment where meaning, data, and action intersect.

If ontologies are commitments about how the world is understood, knowledge graphs surface the consequences of those commitments when data collides with reality. They make explicit what was previously implicit. They expose assumptions and show where interpretations align - and where they break.

This is why knowledge graphs are indispensable to AI governance.

Legacy data architectures are optimized for storage and retrieval. They answer questions like: *Where is the data? Who owns it? What system produced it?* These questions matter, but they do not explain how data is actually used, combined, or interpreted over time.

Knowledge graphs answer different questions.

They show how entities relate across systems. They track how data is derived, inferred, and reused. They capture how meaning propagates as information moves through models, workflows, and decisions. They make traceability navigable rather than forensic.

This shift is subtle but transformative.

In conventional environments, understanding impact requires reconstruction. When something goes wrong, teams piece together lineage, logs, and documentation after the fact. Accountability emerges slowly, if at all. Governance operates reactively.

Knowledge graphs invert this dynamic.

By representing entities, relationships, usage, and context in a connected structure, they allow organizations to see how decisions are formed *as they are formed*. They make it possible to trace not just where data came from, but how it was interpreted, what assumptions were applied, and which downstream outcomes depended on it.

This is where governance becomes operational.

A knowledge graph does not replace judgment. It **creates the conditions under which judgment can be exercised with evidence.** It provides the connective tissue that allows Guardians to enforce boundaries in context, Makers to understand the implications of change, and Seekers to trust - or question - the outputs they receive.

Importantly, knowledge graphs surface consequences.

When ontologies define meaning, they necessarily embed assumptions: about equivalence, relevance, risk, and use. Knowledge graphs reveal how those assumptions behave under pressure. They show where data that appears comparable is not. They show when inference crosses into overreach. They show when context has been lost.

This is not a flaw.

This is the point.

Governance cannot function if assumptions remain hidden. Defensibility depends on visibility - into evidence standards, permitted inference, and stewardship decisions. Knowledge graphs provide that visibility by connecting meaning to action in a way that can be examined, tested, and explained.

This is why knowledge graphs are not simply technical components.

They are **epistemic infrastructure**.

They define how an organization knows what it knows, how it justifies decisions, and how it defends those decisions under scrutiny. When regulators ask why an outcome occurred, when customers challenge fairness, when executives demand explanation, the knowledge graph is where answers are found - not as narratives constructed after the fact, but as traceable paths grounded in evidence.

AI makes this unavoidable.

As models generate inferred attributes, predictions, and recommendations, the distance between input data and output decision grows. Without a knowledge graph, that distance becomes opaque. With one, it becomes navigable.

This does not mean every organization needs a perfect, universal graph.

Effective knowledge graphs grow iteratively. They begin with high-value entities, high-risk relationships, and critical decision paths. They expand as usage expands. Their value compounds because each new connection strengthens understanding across the system.

When knowledge graphs are absent, organizations compensate with restriction.

They limit access broadly. They slow deployment. They centralize approval. These measures reduce exposure, but they also reduce learning and trust. When knowledge graphs are present, organizations can govern precisely rather than bluntly.

This is the difference between control and confidence.

Control attempts to prevent failure by limiting action.
Confidence enables action by making consequences visible.

Knowledge graphs support confidence.

They allow organizations to see not just *what* happened, but *why it happened*, *who was involved*, and *what assumptions were in play*. They make it possible to revisit decisions without rewriting history. They support learning without erasing accountability.

In this way, knowledge graphs complete the system introduced in earlier chapters.

Ontologies define meaning.
Knowledge graphs connect meaning to reality.
Governance emerges from the interaction between the two.

Knowledge graphs do not only organize information. They create the structural conditions required for governance to operate at the point of transaction.

The chapters that follow will show how this connective layer enables data products, marketplaces, and AI systems that are reusable, trustworthy, and scalable - not because risk has been eliminated, but because it can be seen, understood, and managed.

AI does not require perfect knowledge.

It requires defensible knowledge.

Knowledge graphs are how organizations make meaning defensible.

# Chapter 13: Data Products as Governed Interfaces

Once access is centralized, the next question becomes unavoidable.

What exactly is being accessed?

For years, organizations struggled to answer this cleanly. Data was exposed as tables, files, extracts, and feeds – technical artifacts shaped by storage decisions rather than user needs. Governance focused on controlling who could reach those artifacts, not on whether the artifacts themselves were fit for purpose or safe to use.

AI makes this ambiguity untenable.

Models and people alike need **interfaces**, not raw material. They need stable, well-defined ways to ask questions and receive answers, without reinterpreting structure or intent every time. This is where data products emerge.

A data product is not simply a dataset with a name. It is a **governed interface to information**, designed for reuse, accountability, and trust.

It presents data in a form that is meaningful to consumers, while enforcing the constraints required to use it responsibly.

But those constraints do not begin at the product layer.

They begin at the **data object level**.

Every data object – an attribute, a record, an event, a document – must carry, at minimum, a classification of sensitivity. Whether information is public, internal, confidential, regulated, or restricted is not a property of the product that exposes it. It is an intrinsic property of the data itself.

Without object-level classification, governance collapses into guesswork. Data products rely on this foundational tagging to establish access rights, apply masking or filtering, and determine appropriate use. A product cannot responsibly expose information if it does not know the sensitivity of the components it contains. This establishes a clear hierarchy.

At the base are **data objects and data assets**.

These are the raw components of the system. They must be classified, tagged, and governed at rest and in motion. This classification is not optional. It is the minimum condition for any downstream reuse. Without it, security controls become blunt and brittle, applied broadly rather than precisely.

Not every reusable interface must become a formal data product. Between raw assets and formal products lies an important intermediate stage: **Data as a Product.**

This layer shapes data for use. It assembles objects into coherent, reusable views or services. It may be less formal than a fully productized offering, but it is still intentional.

It respects object-level classification and enforces access accordingly. It signals how information is meant to be consumed, even when documentation is light.

Data as a Product is where reuse begins to scale.

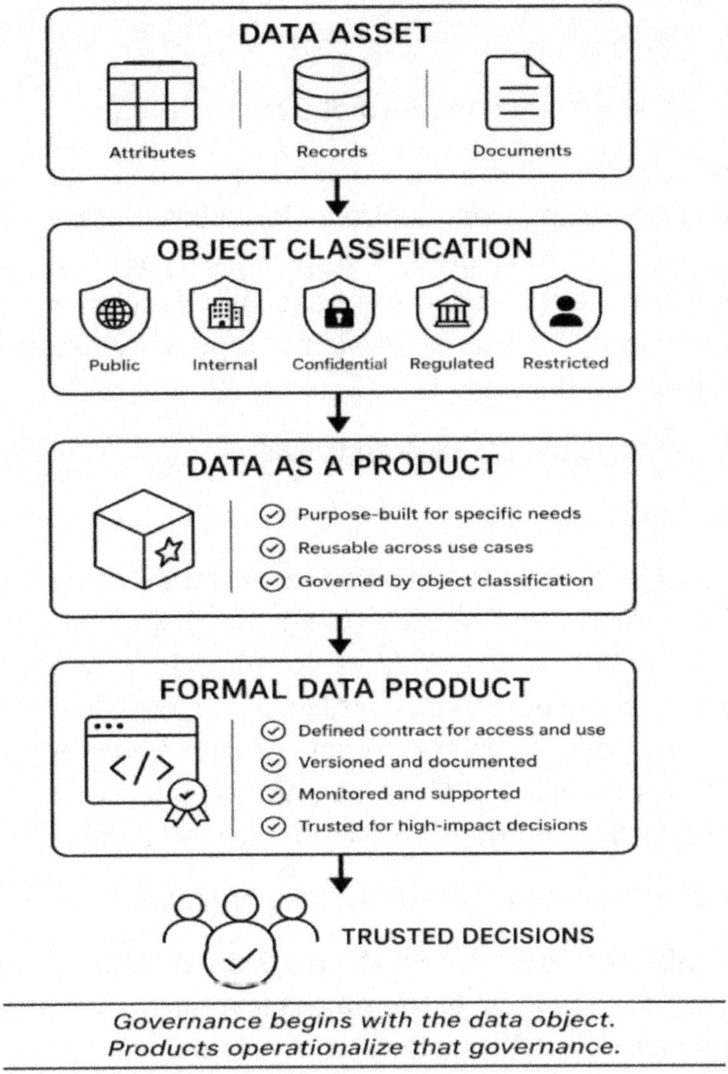


*Figure 3-13-1: Data Objects and Products*

It allows teams to move away from ad hoc extraction while avoiding the overhead of full productization. It creates a bridge between raw assets and formal interfaces, grounded in the security context already established at the object level.

At the top of the spectrum are **formal Data Products**.

These are explicitly designed, documented, versioned, and supported. They define clear contracts for access, usage, and change. They are monitored and governed as first-class interfaces. Formal data products are essential for high-impact, high-risk use cases such as AI systems, regulatory reporting, and core business decisions.

Formal products do not redefine sensitivity.

They **enforce it**.

They apply object-level classification consistently across use cases. They ensure that security rules travel with the data, regardless of how it is combined or consumed. They reduce ambiguity by making governance visible at the interface rather than hidden in backend systems. A data product becomes truly governed when the roles of Guardian, Maker, and Seeker are embedded directly into how it is defined, built, and used.

This layered model changes how security operates.

Instead of enforcing controls only at the system or dataset level, organizations enforce them **at the object level first**, and **at the product level second**. This allows access to be precise rather than restrictive. Sensitive elements can be masked, filtered, or excluded without blocking discovery or reuse entirely.

This approach aligns naturally with activity-based accountability.

Guardians define classification standards and usage boundaries.
Makers assemble and maintain interfaces that respect those boundaries.
Seekers interact with products confidently, knowing sensitivity has already been considered.

Governance becomes embedded rather than reactive.

Importantly, data products do not require a single architectural pattern.

They may be delivered through APIs, query endpoints, views, or services. They may serve humans, models, or both. What unifies them is not their form, but their role as **controlled points of interaction** between classified data and decision-making.

This is what makes them foundational to AI readiness.

AI systems trained or supplied through governed data products inherit object-level security by design. They receive consistent representations with sensitivity preserved. They can be monitored, constrained, and audited over time. When issues arise, responsibility can be traced without ambiguity.

Without object-level classification, governance must rely on coarse restrictions.

With it, data products become powerful, reusable, and safe.

The next chapter extends this model beyond individual interfaces to the ecosystem as a whole – where governed products can be discovered, compared, and reused through marketplaces designed to democratize access without compromising control.

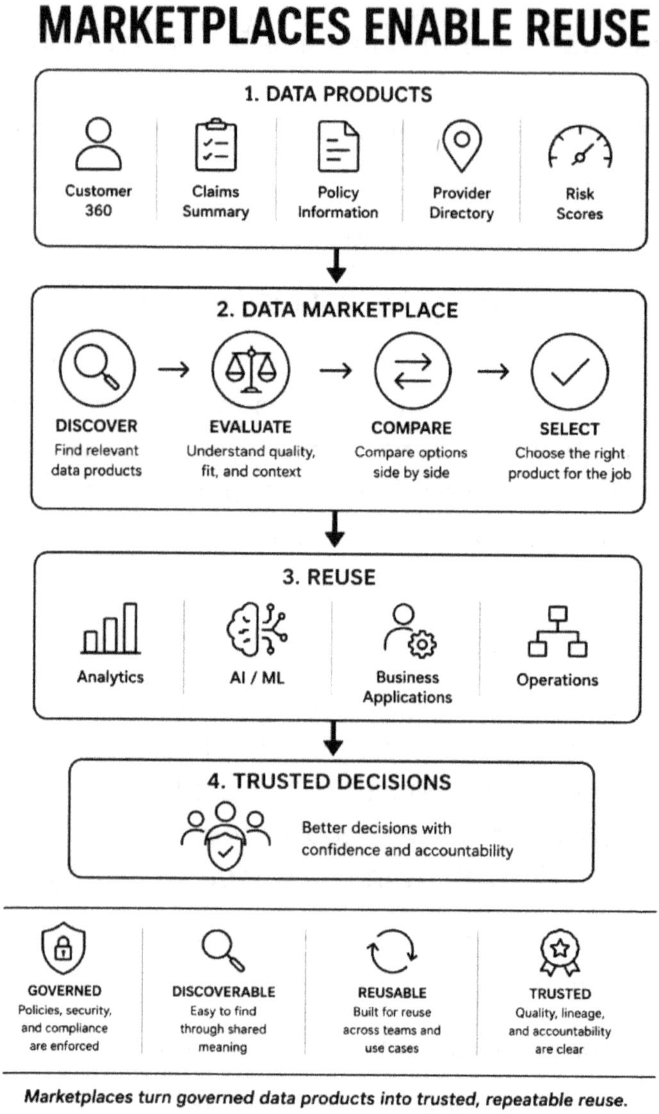


*Figure 3-14-1: Data Marketplaces*

# Chapter 14: Data Marketplaces and Reuse

Once data is exposed through governed interfaces, a new possibility emerges.

Choice.

Instead of asking where data lives or who controls it, consumers begin asking a different question: *Which data should I use to answer this?* This shift marks the transition from managed access to **intentional reuse**.

This is the purpose of a data marketplace.

A data marketplace is not primarily a technical platform. It is a **discovery and decision environment** where data products can be explored, evaluated, and selected based on fitness for use. It allows users to “shop” for information the way they shop for software or services - by understanding what is available, what it represents, and how it can be used.

This metaphor matters. Shopping implies comparison, transparency, and informed choice. The marketplace does not merely expose inventory; it provides context so consumers can decide responsibly. In legacy environments, reuse is accidental Teams copy data because it exists. Models are rebuilt because prior work is hard to find or trust. Governance responds by tightening controls, which further discourages sharing. The result is fragmentation disguised as autonomy.

Marketplaces invert this dynamic.

They encourage reuse by making governed data visible and accessible. They reduce duplication by signaling which products are trusted, supported, and appropriate for specific use cases.

They create incentives for quality because products that are clear and reliable are used more often.

This is where conversational AI becomes transformative.

When users can ask natural language questions:

> *"What customer data can I use for this analysis?"*
>
> *"Which product supports this decision?"*

The barrier to discovery collapses. The marketplace becomes an interface for intent, not just search.

Behind the scenes, this interaction is deeply technical.

Metadata must be accurate. Ontologies must align meaning. Knowledge graphs must connect usage and lineage. Access controls must be enforced dynamically. But none of this complexity is exposed to the user.

That is the point.

The purpose of a marketplace is not to showcase infrastructure. It is to **democratize access to information while preserving control**. It allows more people to ask better questions without increasing risk.

This is where security becomes an enabler rather than an obstacle.

Instead of blocking access broadly, marketplaces apply constraints precisely. Users see what they are allowed to see.

Sensitive products may appear with restricted detail or guided access paths. Governance is present, but not punitive.

This approach changes organizational behavior.

When reuse is easy and safe, teams stop hoarding data. When discovery is intuitive, innovation accelerates. When accountability is embedded, trust grows. The marketplace becomes a learning system, reflecting how information is actually used.

Importantly, marketplaces do not eliminate judgment.

They surface options, not answers. They provide guidance, not guarantees. Seekers remain responsible for interpretation and action, but they are no longer operating in the dark.

This is the culmination of the system described in Part III.

Living maps provide shared understanding.
Ontologies teach machines meaning.
Knowledge graphs make meaning navigable.
Centralized query layers govern access.
Data products define interfaces.
Marketplaces enable reuse.

Together, they create an environment where AI can operate responsibly at scale - not because risk has been eliminated, but because it has been made visible, manageable, and shared.

At scale, however, reuse is only possible when accountability travels with the data. This is what the Guardian, Maker, Seeker model makes possible.

The next part of this book shifts perspective again.

With the system in place, the question is no longer whether AI can be governed. It is whether organizations are prepared -

culturally, operationally, and strategically - to lead in this new environment.

That is the focus of what comes next.

# Part IV - The AI FUTURE

The previous parts of this book established two realities.

- First, the governance models that organizations rely on today were never designed for AI.
- Second, governing AI is not a matter of control alone - it requires systems, roles, and architectures that can operate at speed.

Part IV shifts perspective.

This section is not about technology design. It is about **organizational readiness**. It addresses how leaders, institutions, and operating models must evolve if the systems described earlier are to deliver value rather than risk.

The **AI FUTURE** mnemonic provides a structured way to think about this shift. It does not describe a sequence of projects. It describes a **set of conditions** that must exist simultaneously for AI to function responsibly at scale. These conditions do not replace AI Governance. They define the environment required for the Guardian – Maker – Seeker model to function at scale.

- **A – AI-First Thinking**
  AI is no longer an enhancement layered onto existing

processes. It is a design assumption that reshapes how work is conceived, decisions are made, and value is created.

- **I – Innovation Through Fluid Models**
  Innovation now depends on flexibility in roles, workflows, and authority. Static operating models cannot keep pace with systems that learn, adapt, and recombine information continuously.
- **F – Foundations Aligned to Business Strategy**
  AI readiness is not a technology roadmap. It is a business commitment. Without strategic alignment, AI initiatives fragment into experiments that never scale.
- **U – Underlying Data Maps and Marketplaces**
  AI depends on shared understanding and reusable access to information. Without mapped meaning and discoverable interfaces, intelligence remains brittle and siloed.
- **T – Trust Built Through Performance**
  Trust cannot be mandated through policy alone. In AI systems, trust is earned when outcomes are explainable, repeatable, and defensible under pressure.
- **U – Upskilling Talent and Redefining Roles**
  AI changes what people do, not whether they matter. Organizations must move beyond static job descriptions toward activity-based responsibility and continuous learning.
- **R – Resolving Fear Through Leadership**
  Resistance to AI is not irrational. It reflects legitimate concern about accountability, displacement, and loss of

control. Leadership must address these concerns directly rather than dismiss them.

- **E – Empowering Technology Teams to Architect Success**
  Architecture does not emerge from command-and-control. It emerges when technologists are trusted to design systems that reflect business intent, risk tolerance, and ethical boundaries.

These elements are interdependent.

Organizations that focus on tools without culture fail. Organizations that focus on culture without systems stall. Organizations that pursue innovation without trust lose legitimacy.

The AI FUTURE mnemonic exists to prevent these failures.

The chapter that follows examines each element in turn - not as abstract ideals, but as practical shifts leaders must make to move from experimentation to endurance in an AI-driven world.

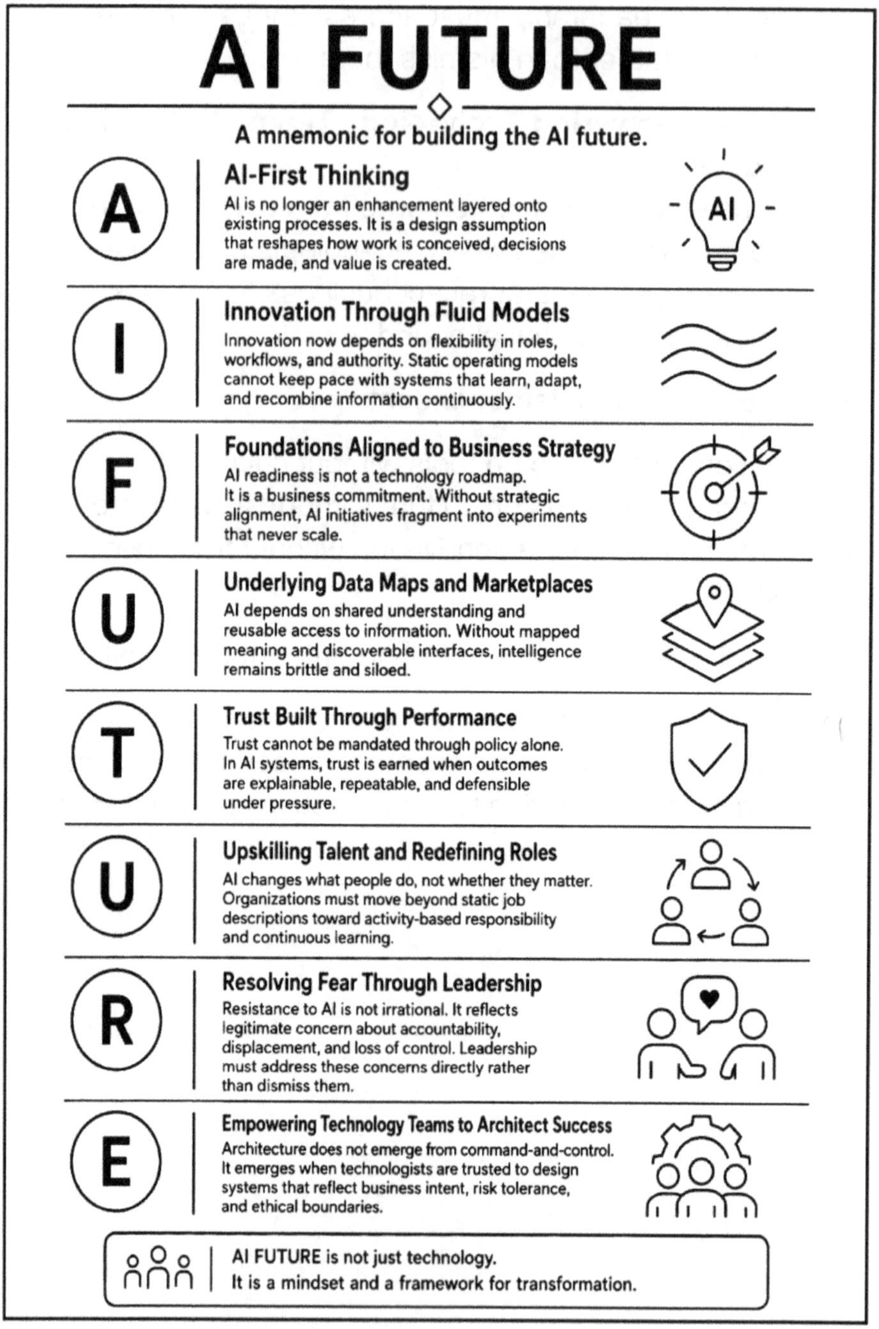


*Figure 4-0-1: The AI FUTURE Mnemonic*

# Chapter 15: AI Future Leadership Perspective

## A = AI-First Thinking

**The Morning Brief:**

The executive did not open a dashboard.

She asked a question.

"What decisions are at risk today?"

The response arrived immediately - not as a chart, not as a report, but as a prioritized narrative. A supplier disruption in one region would cascade impacts into missed delivery commitments within forty-eight hours. A regulatory change introduced overnight created exposure in two customer segments. A model retrained the previous evening had shifted confidence thresholds in a way that warranted review.

Each insight was accompanied by context. Assumptions were visible. Alternatives were suggested, not prescribed. Accountability was clear.

- She did not ask how the system knew this.
- She asked for a prioritized worklist.
- This was not a breakthrough moment. It was routine – just another Tuesday.

The organization had stopped treating AI as a tool and started treating it as part of how decisions were made.

## From AI as a Tool to AI as an Assumption

Most organizations talk about AI as an initiative.

They fund pilots. They stand up centers of excellence. They track use cases. They ask whether AI can improve the process, accelerate workflow, or reduce costs. In this framing, AI is an addition - something layered onto an existing operating model.

That framing is already obsolete.

AI-first thinking does not ask *where AI fits*. It assumes AI is present and asks *how work must change because of it*. It treats intelligence as an ambient capability, not a specialized function. Decisions, processes, and roles are designed with the expectation that AI will participate.

This is not a technical distinction. It is a cognitive one.

Organizations that fail to make this shift experience persistent friction. AI systems exist, but they are underutilized or mistrusted. Outputs are impressive in isolation but disconnected from decision-making. Governance reacts after the fact, attempting to control systems that were never designed to be governed by committee.

AI-first organizations invert this pattern.

They begin with the assumption that AI will influence decisions at scale. They design workflows to accommodate machine input. They define boundaries before deployment rather than after failure. They ask how accountability should function when decisions are assisted, accelerated, or automated.

AI-first thinking changes the questions leaders ask.

Instead of asking: “Can we use AI here?”
The question becomes: “How should intelligence participate in this decision?”

Instead of “Who owns this system?”
The question becomes “Who is accountable for outcomes when AI is involved?”

Instead of “How accurate is the model?”
The question becomes “Under what conditions should we trust the result?”

These questions force clarity.

These questions expose more than technical gaps. They reveal weaknesses in governance, authority, and accountability that were invisible when decisions were made only by people.

AI-first thinking is not something leaders can delegate. It is a leadership philosophy. It shapes strategy, investment, organizational design, and the organization's willingness to rethink long-held assumptions. Whether AI becomes a competitive advantage or a recurring source of disruption is ultimately a leadership decision.

AI-first organizations also redefine failure. The objective is not flawless AI. It is continuous learning with clear accountability. Organizations that demand perfection will constrain AI until it provides little value. Organizations that govern continuous improvement will create systems that become more capable over time.

This requires maturity.

Leaders must accept that AI systems will surface uncomfortable truths. They will challenge intuition. They will reveal inconsistencies in policy and practice. AI-first organizations do not suppress these signals. They design for them.

Most importantly, AI-first thinking shifts the role of humans.

Humans are no longer the sole source of analysis, nor are they passive recipients of machine output. They become supervisors of intent, arbiters of judgment, and stewards of trust. Their value increases as decisions become more complex, not less.

This is the foundation upon which the rest of AI FUTURE is built.

Without AI-first thinking, innovation stalls. Governance becomes reactive. Trust erodes.

With it, the systems described throughout this book become practical. Roles align. Architecture has purpose. Accountability becomes possible.

**AI-first thinking is not about replacing human judgment. It is about redesigning the enterprise so human and machine judgment can operate together at the speed of the business.**

The next section explores what this redesign demands from organizational models themselves, and why rigidity is no longer an option.

## I = Innovation Through Fluid Models

### The Reassignment

The issue did not result in a meeting.

It ended with readjustment.

A risk surfaced during model evaluation - not a flaw, but an edge case that challenged existing policy. No committee was formed. No escalation memo was drafted. Instead, accountability shifted. A Guardian stepped in to define constraints. A Maker adjusted logic and data inputs. A Seeker validated the outcome against lived reality.

By the end of the day, the system behaved differently.

No titles changed.

No org chart was updated.

But responsibility moved to where the work was happening.

Fluid operating models enable organizations to respond at the speed of change rather than the speed of escalation.

## Why Rigid Models Fail at AI Speed

Legacy operating models are built on stability.

Roles are fixed. Decision rights are assigned in advance. Authority flows through formal structures designed to minimize variance. This approach works when change is incremental and predictable.

AI makes change continuous.

Models learn. Data shifts. Context evolves. Decisions that once occurred quarterly now occur in real time. In this environment, rigid operating models become a liability. They introduce delay where speed is required and diffusion of responsibility where clarity is essential.

Innovation stalls not because ideas are lacking, but because authority cannot move.

AI-first organizations confront this reality directly. They recognize that innovation at AI speed requires **fluid models of responsibility** - structures where accountability follows activity rather than hierarchy.

Fluidity does not mean chaos.

It means that roles are activated based on what is happening, not who someone is. A person may act as a Guardian in one moment, a Maker in the next, and a Seeker shortly thereafter. What matters is not the title they hold, but the nature of the decision being made.

This shift is uncomfortable for organizations accustomed to rigid boundaries.

Job descriptions blur. Approval paths shorten. Authority becomes contextual. But this discomfort is a signal that the model is adapting to reality rather than resisting it.

Innovation depends on this adaptation.

When responsibility is fixed but work is dynamic, gaps emerge. Decisions are delayed while accountability is negotiated. Risks go unaddressed because no one is empowered to act. Opportunities are missed because action requires permission from structures designed for a slower world.

Fluid models close these gaps.

They allow organizations to respond in context, assigning responsibility where insight is freshest and impact is greatest. They make it possible to govern change as it occurs, rather than reviewing it after the fact.

This is especially critical in AI systems.

Model behavior can shift based on new data, retraining, or external conditions. Waiting for periodic reviews is insufficient. Governance must be continuous, and that continuity requires operating models that can flex without breaking.

Fluidity also reshapes innovation culture.

Teams stop optimizing for approval and start optimizing for outcomes. Experimentation becomes safer because boundaries are clear and accountability is immediate. Failure is addressed quickly and locally rather than escalated and politicized.

Importantly, fluid models do not eliminate leadership. They demand more of it.

Leaders must define the conditions under which responsibility moves. They must establish guardrails, not micromanage execution. They must trust teams to act within defined boundaries while holding them accountable for results.

This balance is difficult, but necessary.

Organizations that cling to rigid models often attempt to compensate with process. More reviews. More documentation. More committees. These measures create the illusion of control while slowing innovation and obscuring responsibility.

- Fluid models replace illusion with intent.
- They align authority with action.
- They allow innovation to proceed without sacrificing governance.
- They recognize that in AI-driven environments, adaptability is not optional - it is foundational.

This is why innovation and governance are no longer on opposite sides.

In fluid models, governance enables innovation by making responsibility explicit and mobile. Innovation succeeds because it is bounded, not because it is unchecked.

The next section explores how these fluid models must be grounded in something durable - foundations that align AI capability with business strategy rather than allowing experimentation to drift without direction.

## F = Foundations Aligned To Business Strategy

**The Sunset Decision**

The proposal was impressive.

The model performed well in testing. The pilot showed promise. The team had invested months of effort refining features, validating assumptions, and demonstrating potential value.

And yet, the decision was to stop.

Not because the work failed, but because it did not align.

The use case advanced technical capability, but it did not advance the business strategy. It solved a problem the organization had decided not to prioritize. The foundation required to support it would divert resources from initiatives that mattered more.

The team was disappointed. But the organization was stronger for the decision.

AI was not steering strategy. Strategy was steering AI.

## Why Foundations Must Follow Strategy

One of the most common failures in AI adoption is not technical.

It is strategic drift.

Organizations launch AI initiatives because they can, not because they should. Pilots proliferate. Proofs of concept accumulate. Teams become busy demonstrating possibility while losing sight of purpose. Over time, the AI portfolio becomes fragmented - impressive in parts, incoherent as a whole.

This happens when foundations are built without strategic alignment.

AI systems are not lightweight experiments. Even small initiatives require data integration, governance, infrastructure, and talent. These foundations persist long after individual use cases are retired. When they are constructed opportunistically, they constrain future options rather than enabling them.

AI-first organizations reverse this sequence.

They begin with business strategy. They identify the decisions, capabilities, and outcomes that matter most. Only then do they invest in the foundations required to support them. AI initiatives are evaluated not just on performance, but on contribution to strategic goals.

This discipline is essential.

Without it, AI becomes a collection of disconnected efforts competing for attention and resources. With it, AI becomes a force multiplier, reinforcing priorities rather than diluting them.

Foundations aligned to strategy share several characteristics:

- They are designed for reuse across multiple use cases, not optimized for a single model.
- They support the flow of information needed for critical decisions, not just analytical curiosity.
- They embed governance in ways that reflect business risk tolerance rather than generic best practices.

This alignment also clarifies ownership.

When AI foundations are tied to strategy, accountability becomes explicit. Business leaders own outcomes. Technology teams design systems to support those outcomes. Governance functions define boundaries that reflect real-world consequences.

The result is fewer surprises.

AI initiatives that no longer serve strategic goals are retired without drama. New opportunities are evaluated in context. Investment decisions become clearer because they are anchored in purpose rather than enthusiasm.

Alignment does not slow innovation. It gives it focus.

Teams know which problems matter. They understand why certain foundations are prioritized. They are empowered to innovate within a coherent direction rather than chasing disconnected opportunities.

This focus is particularly important as AI capabilities expand.

As models become more powerful, the temptation to apply them everywhere increases. Strategy provides the discipline to resist this temptation. It ensures that intelligence is applied where it creates value and withheld where it introduces unnecessary risk.

Foundations aligned to business strategy also improve resilience.

When market conditions change, strategic priorities shift. Foundations built around enduring capabilities - such as customer understanding, operational resilience, or regulatory compliance - adapt more easily than those built for narrow use cases. The organization can pivot without rebuilding from scratch.

This is the difference between experimentation and endurance.

AI-first organizations are not defined by how many models they deploy. They are defined by how well their systems support the decisions that matter most.

The next section examines one of those foundational capabilities in detail - the underlying data maps and marketplaces that allow intelligence to be shared, discovered, and reused without sacrificing governance.

## U = Underlying Data Maps and Marketplaces

### The Question That Finally Changed

The Chief Operating Officer did not ask for a report.

She asked, "What data can I trust to answer this?"

The pause that followed was telling.

Not because the data did not exist, but because no one could say - with confidence - which version was authoritative, which interpretation was appropriate, or which assumptions were embedded. The organization had information everywhere, but understanding nowhere.

That was the moment the problem became clear.

AI does not fail because data is missing.
AI fails because meaning is fragmented.

## Why Shared Maps Matter More Than Shared Storage

For years, organizations invested in centralizing data.

Warehouses were consolidated. Lakes were standardized. Platforms were unified. These efforts improved efficiency, but they did not solve the core problem that AI exposes.

Centralization of storage does not create shared understanding.

AI systems operate across domains. The underlying data ecosystem became dynamic, distributed, and continuously acted upon. Data is no longer sitting in systems waiting to be analyzed. It is being used, continuously, by models, agents and automated decision pathways. They recombine information in ways no single team anticipates.

When meaning is implicit or localized, every new use requires reinterpretation. Errors compound. Trust erodes.

Shared data maps solve this problem by making meaning explicit across the enterprise.

Information must also be accessible.

This is where marketplaces become essential.

A data marketplace is not a repository. It is a **decision environment**.

It allows users to discover what information exists, understand how it can be used, and select the appropriate data products to answer their questions. It replaces ad hoc sharing with intentional reuse. It shifts behavior from copying data to consuming governed interfaces.

For AI systems, this distinction is critical.

Models trained on inconsistent or unofficial data inherit that inconsistency. Models that draw from trusted, well-described sources behave more predictably. Marketplaces provide the mechanism for establishing and reinforcing that trust at scale.

Importantly, marketplaces do not democratize data indiscriminately. They democratize **visibility** and **choice**.

When meaning and access are shared, responsibility becomes clearer. Decisions can be traced back to the sources and assumptions that informed them. Disagreements can be resolved by examining context rather than debating opinions.

This is how trust becomes operational.

Organizations that lack these capabilities compensate with restriction. They limit access broadly, slow decision-making, and centralize control. These measures reduce risk temporarily, but they also reduce learning and agility.

Organizations with shared maps and marketplaces take a different path.

They allow intelligence to circulate within defined boundaries. They empower teams to act with confidence. As AI systems become more autonomous and pervasive, the cost of fragmented meaning increases. Without shared maps and marketplaces, organizations cannot scale trustworthy intelligence. They only scale uncertainty.

The next section examines how trust itself must evolve in this environment - and why performance, not policy, ultimately determines whether AI systems are accepted or rejected.

# T = Trust Built Through Performance

**The Audit That Wasn't**

The request arrived without warning.

A regulator asked for an explanation – not just of a decision, but of how that decision had been reached, what data informed it, which assumptions were applied, and who was accountable for the outcome.

There was no scramble, no panic, no firefighting.

The response was assembled in hours, not weeks. Evidence paths were clear. Constraints were visible. Decisions were traceable to inputs, logic, and use. What could not be explained was acknowledged openly. What could be defended was defended confidently.

The conversation ended quickly.

Trust was not negotiated. It was demonstrated.

## Why Policy Alone Cannot Create Trust

Most organizations attempt to build trust in AI through documentation.

- Policies are written.
- Principles are published.
- Ethics statements are approved.

These artifacts are important, but they are insufficient. They describe intent, not behavior. When systems operate at speed, trust is shaped by outcomes, not promises.

AI makes this distinction unavoidable.

Models act continuously. Decisions propagate rapidly. When something goes wrong, the question is not whether a policy exists. Instead, the question is whether the organization can explain what happened, correct it, and prevent recurrence.

Trust is therefore earned through performance.

This does not mean accuracy alone.

An AI system can be statistically impressive and still untrustworthy. Bias, opacity, misuse, or misalignment with intent can undermine confidence even when predictions are correct. Trust depends on a broader set of criteria: explainability, consistency, accountability, and responsiveness under pressure. Information must be transparent, understandable, and available

When organizations understand this, trust is explicitly designed.

- They do not rely on post hoc explanations assembled during crises.
- They embed observability, traceability, and governance into systems from the outset.
- They assume scrutiny and prepare for it.

This preparation changes behavior.

Teams build with defensibility in mind. Makers understand how changes will be evaluated. Guardians define constraints that can be enforced dynamically. Seekers receive outputs that carry context rather than blind authority.

When trust is treated as an operational requirement rather than a reputational concern, governance becomes practical.

Performance-based trust also changes how failure is handled.

In low-trust environments, failures trigger blame and retreat. Access is restricted. Innovation slows. In high-trust environments, failures trigger learning. Systems are examined. Assumptions are challenged. Improvements are made transparently.

This distinction determines whether AI adoption accelerates or stalls.

Trust built through performance is cumulative.

Each defensible decision reinforces confidence. Each transparent correction strengthens legitimacy. Over time, stakeholders, regulators, customers, employees, and partners develop an expectation of competence rather than fear of failure.

This expectation matters.

As AI systems become more autonomous, tolerance for error decreases. Stakeholders will not accept opaque systems that cannot explain themselves. They will demand evidence that governance is real, not symbolic.

Organizations that rely on policy alone will struggle to meet this demand. Those that invest in systems and operating models designed for performance-based trust will find themselves ahead of regulation rather than reacting to it.

Importantly, trust is not binary. It is contextual.

An AI system may be trusted for some decisions and not others. Trust must be calibrated based on risk, impact, and use. Performance-based trust allows for this calibration because it exposes the factors that matter.

This is why trust cannot be centralized or standardized fully.

It must be earned repeatedly, in context, through behavior that aligns with stated intent. Systems must support this behavior. Leadership must demand it. A social contract for data is fulfilled when systems consistently demonstrate that boundaries are respected, decisions are explainable, and accountability is visible when intervention is required.

In this sense, trust is not a promise; it is a performance.

Trust is not declared; it is engineered into the moment of execution.

This is why trust can no longer live in policy documents or oversight committees. It must live inside the roles, constraints, and decision pathways that govern action in real time. The Guardian – Maker – Seeker roles are not organizational constructs. They are enforcement mechanisms for trust at speed.

The future of trust is not tied to how much data we have. It is tied to how safely we can act on it at speed.

The section that follows turns toward the human dimension of this shift - how talent, roles, and leadership must evolve to sustain trust without stalling progress.

## U = Upskilling Talent and Redefining Roles

### The Job Description That Didn't Fit

The role had been updated three times in two years.

Each revision added new tools, new expectations, and new responsibilities. None of them captured what the person was actually doing day to day. The work had changed faster than the language used to describe it.

Performance reviews were confusing. Accountability was blurred. Training lagged behind reality.

The problem was not skill. It was structure.

## Why Roles Must Evolve Before Skills Can

Much of the public conversation about AI and work fixates on reskilling.

Organizations debate which technical skills employees must acquire, which roles will disappear, and which new titles should be created. These conversations are understandable, but they often miss the deeper issue.

AI changes how work is done. Skills evolve in response.

- Decisions move faster.
- Boundaries blur.
- Judgment is exercised in new places.
- People interact with AI systems continuously rather than episodically.

In this environment, static job descriptions become misaligned almost immediately.

Leadership's responsibility is not simply to provide training. It is to create an environment where people can evolve as quickly as the work itself. Because AI capabilities and job expectations continue to change, the most durable investment is broad AI and data literacy rather than mastery of any single tool.

Learning focuses less on tools and more on judgment. Employees learn how to assess risk, interpret model behavior, and recognize when escalation is required. Technical literacy increases, but it is contextual rather than generic.

This approach also reduces fear.

When people understand how AI changes their responsibilities - and where their judgment remains essential - uncertainty diminishes. Work becomes clearer, not more precarious. The narrative shifts from replacement to partnership.

Importantly, not everyone needs the same depth of expertise.

Upskilling must be differentiated. Some roles require deep understanding of model behavior and data foundations. Others require fluency in interpretation, questioning, and oversight. Treating all employees as potential data scientists is neither realistic nor necessary.

What is required is **shared literacy**.

Employees must understand enough to participate responsibly. They must know when to trust outputs and when to challenge them. They must recognize the limits of AI assistance and their own accountability within the system.

This shared literacy is foundational to governance.

Without it, policies are ignored or misunderstood. Systems are misused unintentionally. Trust erodes quietly. With it, governance becomes part of everyday work rather than an external imposition.

Redefining roles inevitably changes leadership expectations, but it does not require a wholesale redefinition of every job. What it requires is a shift in how performance is understood and evaluated. Managers must look beyond task completion and assess the quality of decisions made in AI-assisted environments. Leaders must create conditions where questioning AI output is expected and valued, not discouraged. Accountability must be reinforced through consistent behavior rather than formal rhetoric.

Job descriptions should evolve at the point of evaluation, updated in partnership with employees to reflect how work is actually being performed. In this way, role clarity becomes an outcome of lived experience, not a preemptive HR exercise.

Upskilling in this context is continuous.

As AI systems evolve, so do the activities they influence. Change is not a one-time event. It is an ongoing adaptation that mirrors the systems it supports.

Organizations that treat upskilling as episodic will fall behind. Organizations that treat it as integral to role evolution will build resilience.

The next section addresses the emotional dimension of this transition - the fear that often accompanies change, and the leadership responsibility to address it directly rather than dismiss it.

## R = Resolving Fear Through Leadership

### The Question No One Asked Out Loud

The silence came after the presentation.

The strategy was sound. The systems were impressive. The roadmap was ambitious but credible. And yet, no one spoke.

Finally, someone asked the question everyone else had avoided.

"What happens to us?"

It was not a technical question. It was not even an organizational one. It was a human question. And it was the right one.

## Why Fear Is a Rational Response to AI

Fear is often treated as resistance.

Leaders interpret hesitation as unwillingness to change, lack of vision, or attachment to outdated ways of working. In the context of AI, this interpretation is not only wrong - it is dangerous.

Fear is a rational response to uncertainty.

AI alters how decisions are made, how work is evaluated, and how accountability is assigned. It challenges professional identity. It exposes performance in new ways. It shifts authority from familiarity to evidence. People are right to be cautious.

Organizations that dismiss fear create *shadow resistance.*

- Concerns go underground.
- Adoption becomes performative.
- People comply publicly while avoiding reliance on systems they do not trust.
- Innovation slows quietly, not loudly.

Resolving fear requires leadership, not persuasion.

## What Leadership Must Make Explicit

Fear thrives in ambiguity.

When people do not know how decisions will be made, how mistakes will be handled, or how performance will be judged, they assume the worst. AI amplifies this effect because its behavior can feel opaque and impersonal.

Leadership must therefore make several things explicit.

**First, accountability.**

People need to know where accountability and responsibility sit when AI is involved. Who is accountable for outcomes. Who can intervene. Who has authority to stop or change a system. What specific role are they allowed or expected to perform. Ambiguity here creates anxiety and avoidance.

**Second, boundaries.**

Employees must understand what AI will and will not be used for. Vague assurances are insufficient. Clear constraints provide psychological safety. When people know where lines exist, they can operate confidently within them.

**Third, expectations.**

AI changes how work is evaluated. Judgment, critical thinking, reliability, and agility are the new critical measures for evaluation. Leaders must articulate what good judgment looks like in an AI-assisted environment. Blind reliance is not rewarded. Thoughtful challenge is. Learning from error is expected.

These signals resolve fear more effectively than any town hall or training session.

## Leadership Behavior Matters More Than Vision Statements

Employees watch what leaders do.

- They observe how leaders respond when AI produces uncomfortable insights.
- They notice whether leaders override systems selectively or accept evidence consistently.
- They see how mistakes are handled and whether accountability is fair.

These behaviors define the real policy – the **culture** that will govern outcomes.

- When leaders blame individuals for system failures, fear increases. When leaders accept responsibility for design flaws and invest in improvement, trust grows.
- When leaders use AI outputs opportunistically, skepticism spreads. When leaders subject themselves to the same governance as everyone else, legitimacy follows.

Resolving fear requires alignment between words and actions.

This alignment is tested under pressure.

- When AI recommendations conflict with intuition
- When models expose inefficiencies or bias
- When decisions must be defended externally.

Leaders who retreat to hierarchy in these moments undermine everything they claim to support. Leaders who engage transparently strengthen the system.

## Pride Replaces Fear When Purpose Is Clear

Fear diminishes when people understand why change matters.

Not abstractly, but concretely.

When employees see how AI improves safety, fairness, resilience, or effectiveness, they begin to take pride in participating. When they understand how their judgment contributes to outcomes, their role feels elevated rather than diminished.

Leadership plays a critical role in connecting AI initiatives to purpose.

This is not about selling a future vision. It is about demonstrating present value. Small, defensible wins matter more than grand promises. Each successful deployment that respects boundaries and delivers benefit builds confidence.

Over time, fear gives way to ownership.

People stop asking whether AI threatens their role and start asking how they can use AI to make their role better. This shift cannot be mandated; it must be earned.

## Resolving Fear Is Ongoing Work

Fear does not disappear permanently.

As systems evolve, new concerns emerge. As roles shift, new uncertainties arise. Leadership must treat fear as a signal, not a failure. It indicates where clarity is lacking or trust has not yet been established.

Organizations that listen to these signals adapt faster.

They refine governance. They adjust training. They revisit boundaries. They demonstrate responsiveness rather than rigidity.

Responsiveness is a critical leadership competency. It can be learned, but it must be expected.

In AI-first organizations, leadership is measured not by certainty, but by the ability to guide others through uncertainty without losing integrity.

The final section turns to the technology teams tasked with building these systems - and the conditions leadership must create for them to architect success rather than fight constraints.

# E – Empowering Technology Teams to Architect Success

### The Constraint That Broke the System

The architecture review failed before it began.

Not because the design was wrong, but because design was never truly allowed. The most consequential decisions had already been made elsewhere - priorities locked, platforms selected, timelines fixed.

Technology teams were expected to plan delivery, but not to shape the design. They were asked to deliver outcomes they did not define, using tools they did not choose, under constraints they could not change, while being expected to resolve problems they were explicitly prevented from addressing.

When the process could not accommodate new architectural insight, responsibility for outcomes did not reflect where the decisions were made - it was pushed downward, to those closest to the work.

This was not a failure of technology teams to deliver. It was a failure of how authority and responsibility were structured.

The pattern was familiar. And it was unsustainable.

## Why Architecture Cannot Be Commanded

Organizations often treat architecture as an execution function.

Leaders define strategy. Governance defines rules. Procurement selects platforms. Technology teams are expected to assemble these inputs into coherent systems. When results disappoint, the response is more oversight, more approval gates, and tighter control.

This approach misunderstands the nature of architecture.

Architecture is not compliance – it is synthesis.

Architects must reconcile competing objectives:

- performance and control
- flexibility and safety
- innovation and stability

These tradeoffs cannot be resolved through mandates alone. They require judgment exercised in context.

AI intensifies this requirement.

AI systems cross traditional data and legacy governance boundaries. They touch data, models, workflows, and decisions simultaneously. Architectural choices have ethical, operational, and regulatory consequences. Treating architecture as an afterthought guarantees misalignment.

Empowering technology teams is therefore not a cultural preference. It is a structural necessity.

## What Empowerment Actually Means

Empowerment does not mean autonomy without accountability.

It means **clear intent, defined boundaries, and trust in execution**.

Technology teams must understand the business outcomes they are responsible for supporting. They must know which risks matter most and where flexibility exists. They must have authority to design within those constraints rather than working around them.

This clarity enables better systems.

When architects are involved early, governance becomes design input rather than post hoc enforcement. When engineers understand why constraints exist, they implement them more effectively. When teams are trusted to make tradeoffs, systems become more resilient.

Empowerment also accelerates learning.

AI systems evolve. Assumptions change. New risks emerge. Teams closest to the system are best positioned to detect these changes. When they are empowered to act, issues are addressed quickly. When they are constrained by rigid approval structures, problems fester.

This difference determines whether AI systems improve or stagnate.

## The Cost of Disempowerment

When technology teams are not empowered, organizations pay a hidden tax.

Systems become brittle because they are optimized for compliance rather than adaptability. Innovation slows because every change requires negotiation. Governance becomes symbolic because it cannot keep pace with reality.

Most damaging of all, trust erodes.

Technology teams disengage. They stop advocating for better designs because effort is not rewarded. They focus on delivery rather than quality. Over time, the organization loses the very expertise it depends on to manage complexity.

*Empowerment is a retention strategy as much as a technical one.*

Leadership's Role in Architectural Success

Leaders do not need to design systems.

They need to create the conditions under which good systems can be designed.

This includes:

- Defining non-negotiable principles
- Articulating acceptable risk
- Aligning incentives with long-term outcomes
- Holding teams accountable for results

When these conditions are met, great architecture emerges naturally.

Systems reflect business intent. Governance is embedded. AI becomes scalable rather than fragile. Technology teams take pride in their work because they are solving meaningful problems, not just implementing directives.

This is how organizations move from experimentation to endurance.

# Part IV Final Thoughts - The AI FUTURE

**The AI FUTURE is not a roadmap.**

**It is a readiness model.**

Each element describes a condition that must exist if AI is to operate responsibly at scale. None of them succeed in isolation. Together, they describe a shift in how organizations think, lead, and build.

- **AI-First Thinking** reframes intelligence as an assumption, not an add-on
- **Innovation Through Fluid Models** aligns responsibility with activity
- **Foundations Aligned to Business Strategy** prevent drift and fragmentation
- **Underlying Data Maps and Marketplaces** create shared understanding and reuse
- **Trust Built Through Performance** replaces promises with evidence
- **Upskilling Talent and Redefining Roles** keeps humans central to judgment
- **Resolving Fear Through Leadership** acknowledges uncertainty rather than denying it
- **Empowering Technology Teams** turns intent into durable systems

This model exists because AI changes the shape of organizations.

It accelerates decisions. It amplifies consequences. It exposes assumptions. It forces clarity where ambiguity once sufficed.

- Organizations that treat AI as a tool will struggle.
- Organizations that treat it as an operating condition will thrive.

The choice is not whether to adopt AI.

The choice is whether to redesign for it.

The next section of this book turns from readiness to execution. It introduces the governance framework that makes these principles operational - the mechanisms by which responsibility, authority, and accountability are enforced in real systems.

Vision alone is not enough.

The future belongs to those who can build it - and govern it - at the same time.

# Part V – Governing AI in Practice

Parts I through IV established a single, unavoidable conclusion:

> **AI did not expose a gap in technology.**
> **It exposed a gap in governance.**

We have seen how intelligence now operates continuously rather than episodically, how decisions unfold across humans and machines, and how responsibility shifts faster than legacy structures can follow. We examined why paperwork-based governance failed, why static roles collapse under AI speed, and why trust, architecture, leadership, and culture can no longer be treated as separate concerns.

By the end of Part IV, the conditions were clear.

Organizations must be able to:

- Govern decisions as they happen, not after the fact
- Align accountability with activity rather than hierarchy
- Preserve trust under scrutiny
- Enable innovation without losing control
- Do all of this without freezing the organization in process

What remained unresolved was **how**.

The AI FUTURE described what must be true for AI to function responsibly at scale. It articulated readiness conditions – strategic, organizational, architectural, and human. What it did not yet provide was an operational governance model capable of sustaining those conditions under real-world pressure.

Part V closes that gap.

From Readiness to Governance That Operates

Governance has traditionally been treated as a layer applied after systems are designed.

In AI environments, that approach fails invisibly until it fails publicly.

When AI participates directly in decisions, governance must operate at the same tempo. When responsibility moves as work unfolds, accountability cannot remain fixed to titles or committees. When systems adapt continuously, governance cannot rely solely on periodic review.

Effective AI governance must function **at the point of action**.

It must be able to answer, in real time:

- Who is accountable for this decision, right now?
- What authority applies in this context?
- What constraints must be enforced before action proceeds?
- How will this decision be defended if challenged?

These questions cannot be resolved through static role definitions or document-based controls. They require a governance model that reflects how work actually happens in AI-enabled organizations.

***Governance*** has moved from **organizational design** to **computational design**.

This is where the EZ-AI-Ready Guardian–Maker–Seeker (GMS) Framework enters. It can be understood as the operational architecture of the new social contract for data in AI environments. The model presented here is not theoretical. It is designed for environments where governance must operate inside execution – at speed, under pressure, and without the ability to pause for review.

## What is the EZ-AI-Ready GMS Framework?

The EZ-AI-Ready Guardian–Maker–Seeker (GMS) Framework is a **patent-pending utility framework** designed to do more than suggest new lines of authority and responsibility. It formally recognizes the transition to **role-based authority**, where accountability is determined by action rather than title. It uses the roles of Guardian – Maker – Seeker as defined in Part II.

The framework is not a product to purchase, nor is it a vague system theory. It is a **governance operating pattern** designed specifically for AI-enabled environments - one that builds trust in people and enforces accountability through behavior.

It functions simultaneously as:

- a **behavioral shift**, making responsibility explicit at the moment of action,

- a **cultural shift**, redefining accountability as situational rather than positional,
- a **realignment of duties**, based on the work being performed rather than the role assigned, and
- an **architectural pattern**, intended to be embedded into systems, workflows, and decision processes going forward.

The framework starts from a simple observation:

> In practice, people already move between different governance capacities throughout the day. Governance fails when organizations ignore this reality and attempt to force dynamic work into static structures.

The EZ-AI-Ready GMS Framework formalizes these patterns so they can be governed, enforced, and defended.

## Guardian – Maker – Seeker Made Explicit

We explored these roles previously as general governance structures. Within this system they become the center of the framework as three **states of responsibility** that exist in every AI-enabled organization:

- **Guardian** – defining constraints, boundaries, and conditions for use
- **Maker** – changing systems, data, models, or logic
- **Seeker** – consuming outputs to make decisions and take action

These are not job titles, teams, or reporting lines.

A single individual may move between these states repeatedly as work unfolds. Governance becomes possible when authority and accountability follow those transitions rather than remaining tied to organizational charts.

This is the core insight the framework operationalizes.

What the Framework Enables

By making these states explicit and governable, the EZ-AI-Ready GMS Framework enables organizations to:

- enforce governance continuously rather than episodically,
- embed constraints into systems instead of documents,
- make decisions traceable, explainable, and defensible,
- move faster without becoming reckless,
- scale AI without scaling risk.

This is how the loose ends from earlier sections are resolved.

The failures described in Part I gain a structural answer.
The role shifts described in Part II become enforceable.
The systems in Part III gain a governance backbone.
The leadership conditions in Part IV become executable.

What This Section Will and Will Not Do

Part V will not revisit readiness, vision, or culture.

It will focus on **governance in practice**.

The chapters that follow will:

- ground the framework in concrete role definitions and enforcement logic,
- show how governance integrates with AI systems and data architecture,

- demonstrate how accountability holds under pressure,
- provide a practical model that organizations can adopt incrementally.

The first chapter of this section will describe the framework as it exists in the formal patent submission – focusing on mechanics and role redefinition as required. Subsequent chapters will expand outward to show how the framework operates across behavior, culture, and systems.

This is not an abstract theory.

It is a practical response to the governance failures AI has already made visible.

The future of AI will be shaped not by who builds the most powerful models, but by who can align responsibility, authority, and action at the speed intelligence now demands.

The EZ-AI-Ready Guardian–Maker–Seeker Framework exists to do exactly that.

# Chapter 16: The EZ-AI-Ready GMS Framework – How It Works

The Guardian – Maker – Seeker Framework is easiest to misunderstand when it is described only at a conceptual level.

At that altitude, it can sound like a philosophy of governance, a cultural aspiration, or a reframing of roles. Those elements are real, but they are not sufficient. What distinguishes the EZ-AI-Ready GMS Framework* is that it is **operational**.

**EZ-AI-Ready GMS Framework is patent pending as of Feb. 2025*

It works because it is anchored in architecture.

This chapter provides an overview of the **technical capabilities** described in the patent – not in patent language, but in terms leaders, architects, and governance teams can recognize and evaluate. The goal is not to teach implementation, but to make clear why this framework is enforceable rather than symbolic.

## Governance as an Architectural Concern

Legacy governance models assume that oversight happens outside the system.

Policies are written. Reviews are scheduled. Committees evaluate outcomes after decisions are made. Enforcement is procedural rather than structural. This model depends on time, hierarchy, and human intervention – all of which AI compresses or bypasses.

The EZ-AI-Ready GMS Framework treats governance differently.

Governance is not layered on top of systems.
It is **designed into them**.

At a high level, the framework introduces a set of architectural capabilities that allow responsibility, authority, and accountability to be evaluated **at the moment of action**. These capabilities are independent of any single technology stack, vendor, or platform. They describe *how* governance must function, not *what* tools must be used.

## Core Technical Capabilities (Conceptual Overview)

The framework relies on several foundational capabilities that work together.

First, **activity-aware role evaluation**.

Rather than assigning authority permanently based on job title, the framework evaluates which state of responsibility is active based on what action is being performed. Changing data, modifying logic, or consuming outputs are treated differently, even if performed by the same individual. Governance follows activity, not identity alone.

Second, **context-sensitive constraint enforcement**.

Constraints are not generic rules applied uniformly. They are evaluated based on context – the data involved, the purpose of use, the risk profile, and the active role state. This allows governance to be precise without being brittle.

Third, **decision traceability by design**.

Actions taken within the system produce a traceable record of:

- who acted,
- in what role state,
- under which constraints,
- using which inputs,
- for what purpose.

This traceability is not an audit artifact assembled later. It is a byproduct of normal operation.

Fourth, **separation of decision, modification, and use**.

The framework enforces a clear distinction between those who define constraints, those who change systems, and those who use outputs. This separation does not slow work. It prevents silent conflicts of interest and enables defensibility when outcomes are challenged.

Finally, **continuous governance, not episodic review**.

Because role state, constraints, and context are evaluated dynamically, governance operates continuously. Decisions do not wait for quarterly reviews. Risk does not accumulate unnoticed. Accountability is immediate and situational.

These capabilities are what allow the framework to scale.

## Why This Is Not Just Another Control Model

It is important to understand what these capabilities do *not* imply.

They do not require centralized approval for every action.
They do not assume perfect data or perfect models.
They do not replace human judgment with automation.

Instead, they create a **governable environment** where judgment – human and machine – can operate without ambiguity.

This is why the framework resonates with architects.

Architects recognize that systems fail when responsibility is implicit, constraints are external, and accountability is retroactive. The EZ-AI-Ready GMS Framework addresses these failure modes structurally rather than procedurally.

It aligns with how modern distributed systems already function – by embedding control points, enforcing boundaries, and recording state transitions as part of normal operation.

Patent Language vs. Narrative Language

At this point, it is worth naming the difference between the formal patent submission and the narration you are reading now.

The patent describes **mechanics**.

It defines role states, enforcement logic, architectural relationships, and system behaviors with precision. It is intentionally narrow, technical, and constrained. Its purpose is to establish novelty, scope, and enforceability.

This book describes **meaning**.

It explores why those mechanics matter, how they change organizational behavior, and what it feels like to operate in systems designed this way. It uses narrative to surface consequences that technical documents cannot convey.

Both are necessary.

The patent ensures the framework can be implemented and defended.

The narrative ensures it can be understood, adopted, and sustained.

What Comes Next

This chapter described a patentable framework.

The chapters that follow will build from here – not by repeating technical detail, but by showing how these capabilities manifest across governance decisions, AI systems, data architectures, and organizational life.

We will return to Guardian – Maker – Seekers – not as abstractions, but as active participants in a system that finally aligns responsibility with action.

The cracked egg is no longer a metaphor alone.

It is now an architecture.

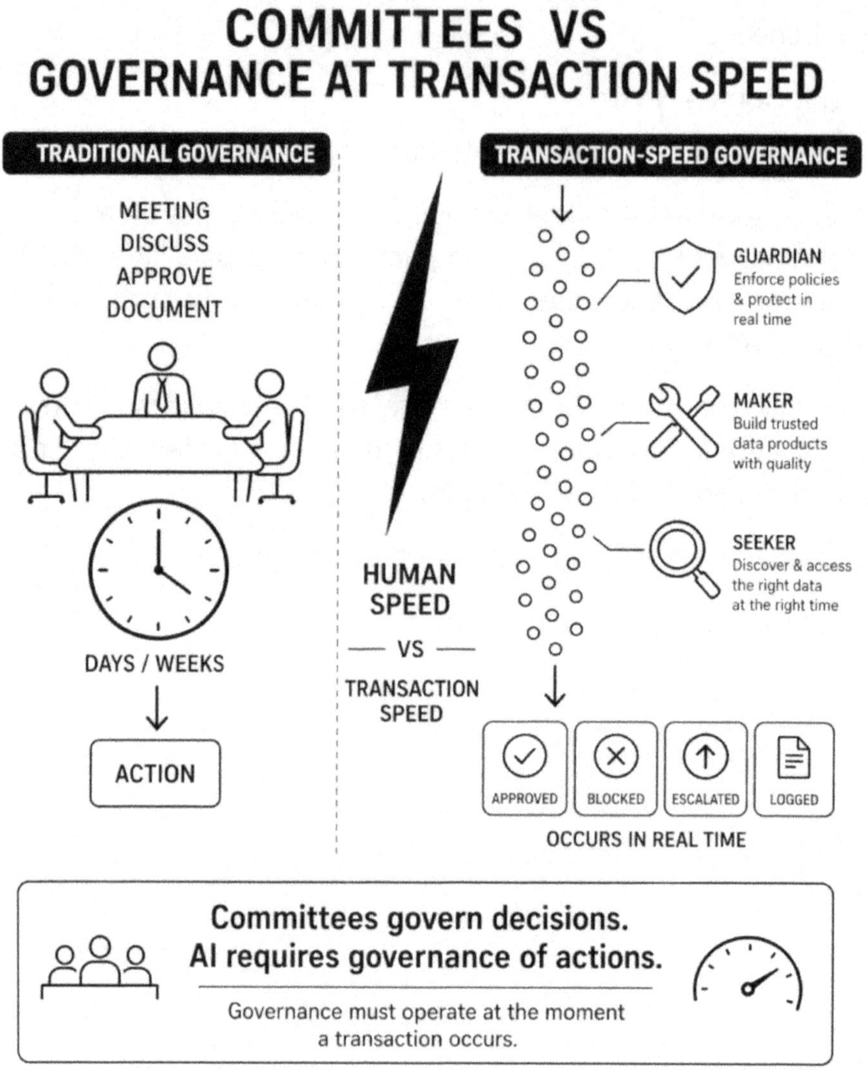


*Figure 5-17-1: Committees vs Governance at Transactional Speed*

# Chapter 17: The Committees

## Committees and Governance at AI Speed

Before describing the Guardian role, it is necessary to address a question many readers will already be holding:

**Where do committees fit now?**

This is not a trivial concern. For decades, committees have been the primary mechanism through which organizations governed data, technology, risk, and ethics. They created alignment across silos, established shared standards, and provided legitimacy for difficult decisions. In many organizations, they still do.

The problem is not that committees are wrong.

The problem is that AI changed **when** governance **must** occur.

## Where Committees Excel

Committees excel at governance activities that are:

- deliberative
- cross-functional
- policy-oriented
- episodic by nature

They are well-suited to:

- defining standards and principles
- establishing risk tolerance and escalation thresholds

- approving frameworks and guardrails
- resolving disputes that require consensus
- reviewing outcomes after sufficient evidence has accumulated

In legacy data and technology environments, this timing worked. Decisions were slower. Systems were more static. Reviews could occur after the fact without significant harm.

Committees brought legitimacy, not speed.

And legitimacy mattered most.

## Where Committees Break Down

AI introduces a different operating reality.

Decisions now occur continuously. Systems adapt in near real time. Risk emerges at the intersection of data, models, and use - often far downstream from where policy was written.

In this environment, committees struggle not because they are ineffective, but because they are **misplaced in the flow of work**.

Committees cannot:

- evaluate every decision as it happens
- arbitrate responsibility at machine speed
- remain continuously present at points of action
- enforce constraints dynamically within systems

When organizations attempt to use committees this way, two things happen.

First, innovation slows. Work queues form around approvals. Teams route decisions upward defensively. Governance becomes a bottleneck rather than a safeguard.

Second, governance is bypassed. Faced with pressure to deliver, teams work around formal processes. Shadow decisions emerge. Risk accumulates silently until it surfaces publicly.

Neither outcome reflects a failure of intent.

They reflect a mismatch between **governance structure and operational tempo**.

## The Shift Committees Must Make

The introduction of the GMS Framework does not eliminate committees.

It **relieves them of duties they were never designed to perform**.

Under the EZ-AI-Ready GMS model, committees move upstream and outward.

They focus on:

- defining the constraints Guardians are responsible for enforcing
- setting policy, standards, and escalation criteria
- reviewing patterns and outcomes rather than individual actions
- adjudicating conflicts that exceed role-level authority
- evolving governance in response to observed system behavior

In other words, committees govern the **system of governance**, not each decision within it.

This is a critical distinction.

Committees remain authoritative, but they are no longer asked to be omnipresent.

How GMS Complements Committees

The Guardian–Maker–Seeker Framework operates where committees cannot - at the moment of action.

GMS roles:

- enforce committee-defined constraints in context
- exercise judgment within approved boundaries
- surface exceptions and signals for escalation
- make governance continuous rather than episodic

Committees and GMS roles are not competing structures.

They are **different layers of the same governance system**.

Committees define intent.
GMS roles execute it.

Committees decide *what must be true.*
GMS ensures it *remains true under pressure*.

## Why This Distinction Matters

Without this clarity, organizations fall into one of two traps.

They either overload committees, asking them to govern work they cannot see in time - or they abandon committees altogether in the name of speed.

Both are mistakes.

The GMS Framework exists precisely to avoid this false choice.

It preserves institutional oversight while enabling operational governance. It honors the role committees have played, while acknowledging the limits imposed by AI-driven environments. Committees do not disappear in an AI-driven organization. Their role changes – from reviewing decisions after the fact to defining the constraints under which decisions are made.

Once this distinction is clear, the Guardian role becomes easier to understand.

Guardians are not committees.
They do not replace committees.
They operationalize the boundaries committees set.

And they do so continuously, in context, and with accountability.

That is where we turn next.

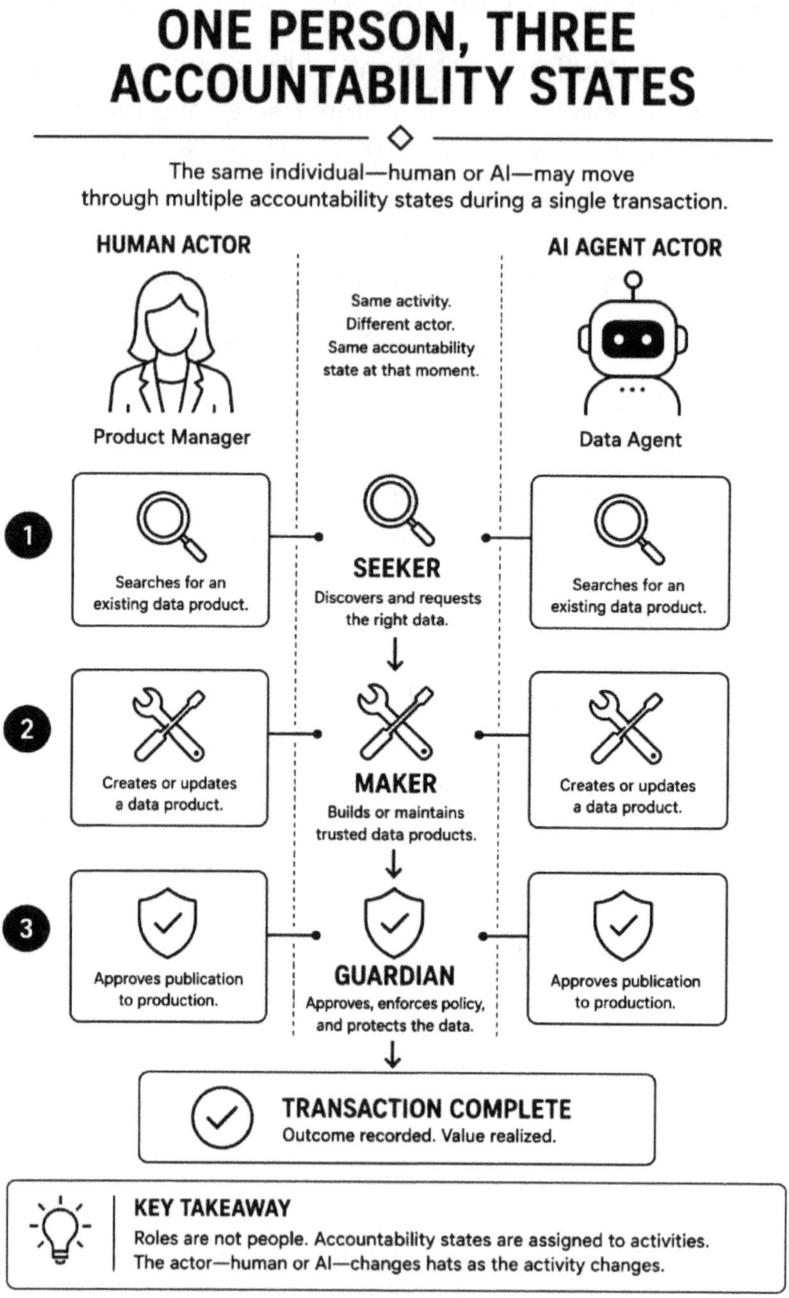


*Figure 5-18-1: Accountability States*

# Chapter 18: From Data Owner to Guardian

Holding the Line Without Slowing the System

The Guardian role is often misunderstood before it is ever defined.

In many organizations, the closest existing analogue is the **Data Owner** – the individual accountable for the appropriate use, protection, and stewardship of information. That accountability remains essential. What changes in AI-enabled environments is **how** it is exercised.

The Guardian role is the evolution of Data Ownership.

Where the legacy Data Owner model relied on static approvals and periodic oversight, the Guardian role operates **continuously**, enforcing boundaries at the moment decisions are made. It replaces episodic control with persistent, contextual governance.

A Guardian is not a gatekeeper.

A Guardian is responsible for **making speed safe**.

## Why Guardians Are Necessary Now

AI changes not only how decisions are made, but **who or what makes them**.

In legacy systems, decisions were assumed to be human, discrete, and reviewable after the fact. Risk could be managed through process and escalation because actions unfolded slowly and predictably.

AI collapses those assumptions.

Decisions now occur continuously. Some are made by humans assisted by models. Others are made directly by **AI agents, automated rules, or system logic** operating within predefined boundaries. Risk emerges not at deployment, but at use.

In this environment, governance that relies on after-the-fact review is no longer sufficient.

Guardians exist because **constraints must be defined and enforced before decisions occur**, regardless of whether the actor is human or machine.

## Guardians and Decision Classes

A core responsibility of the Guardian role is defining **decision classes**.

A decision class groups decisions based on shared characteristics, such as:

- potential impact
- sensitivity of the data involved
- reversibility of outcomes
- regulatory or contractual exposure
- and tolerance for automation

Not all decisions require the same level of oversight.

Some decision classes may be:

- fully automated
- delegated to AI agents within strict bounds
- executed by humans with AI assistance

- or reserved for explicit human judgment and escalation

Guardians define these boundaries.

They determine **which decisions may be automated**, under what conditions, and using which data. They specify when human review is required, when escalation must occur, and when execution must be blocked entirely.

This is how governance becomes precise rather than blunt.

## Security Profiles and Enforcement

Decision classes are inseparable from **security profiles**.

Guardians are responsible for ensuring that decisions – whether made by humans or machines – are constrained by appropriate access rights, data classifications, and use conditions. This includes:

- who or what may access specific data
- for what purpose
- at what level of aggregation
- and under what contextual conditions

AI agents do not receive blanket authority.

They operate within **explicitly defined security profiles** that reflect the same governance principles applied to human actors. The difference is not in rigor, but in enforcement.

When constraints are embedded into systems, governance is no longer dependent on perfect behavior. It is enforced by design.

## What a Guardian Actually Does

The Guardian role is responsible for defining and enforcing the **conditions under which action is permitted**, regardless of the actor.

This includes:

- establishing acceptable use boundaries
- defining decision classes and automation limits
- specifying required context for use
- determining escalation thresholds
- ensuring constraints are applied consistently across systems

Guardians do not approve every action.

They define the **shape of the decision space**.

By doing so, they enable Makers and Seekers – human and non-human alike – to operate decisively within known boundaries. Speed is preserved because ambiguity is reduced.

**Guardians Are Not Committees**

Guardians are not standing committees, review boards, or approval gates.

Committees define intent.
Guardians operationalize it.

Committees decide what must be true at a policy level.
Guardians ensure those conditions hold **in context**, at the moment a decision is executed.

This distinction matters even more when AI agents are involved. Committees cannot govern machine-speed decisions. Guardians can – by designing the constraints that machines are allowed to operate within.

Guardianship Is a State of Responsibility

Like all GMS roles, Guardian is a **state of responsibility**, not a permanent title.

A person may act as a Guardian while defining constraints, then transition into a Maker role to implement them, or into a Seeker role when evaluating outcomes. What matters is that **only one state is active at a time**, and accountability follows that state.

AI agents may also operate within Guardian-defined boundaries, but they do not assume Guardian responsibility themselves. Accountability remains human, even when execution is automated.

This preserves responsibility without sacrificing speed.

**How Guardians Enable Trust**

Trust in AI systems is not built through assurances.

It is built through **predictable behavior under constraint**.

Guardians make trust possible by ensuring that:

- boundaries are explicit,
- automation is intentional,
- exceptions are visible,
- and accountability is unambiguous.

When decisions are challenged – by regulators, customers, or internal stakeholders – Guardian-defined constraints make defensibility possible. Outcomes can be explained in terms of intent, decision class, and permitted inference rather than justification after the fact.

## What Guardians Do Not Do

Guardians do not approve every decision.
They do not manually oversee every AI action.
They do not eliminate the need for human judgment.

They design the conditions under which judgment – human or machine – can be exercised responsibly.

Their success is measured not by how often they intervene, but by how rarely intervention is required.

### From Data Ownership to Guardianship

The Guardian role does not discard the concept of Data Ownership.

It completes it.

Where Data Owners were responsible for stewardship in static environments, Guardians are responsible for **operational stewardship in dynamic, AI-enabled systems**. They hold the line not by slowing work, but by shaping the space in which work occurs.

This is how governance keeps pace with intelligence.

# Chapter 19: From Data Steward to Maker

## Changing Systems Without Losing Control

The Guardian defines the boundaries of action; the Maker operates within them.

In most organizations, the role most closely associated with the Maker is the Data Steward – the individual responsible for maintaining data quality, consistency, and correctness. That responsibility remains essential. What changes in AI-enabled environments is the scope and visibility of the work.

The Maker role expands beyond stewardship of static data.

Makers are anyone who changes something in the system.

They modify data, adjust logic, configure rules, retrain models, enrich records, correct errors, and execute transactions. This includes legacy data stewards, engineers, analysts, and product teams – but also frontline staff performing operational or transactional work.

If an action alters the state of the system, it is a Maker action.

### Why the Maker Role Matters More Than Ever

In legacy environments, changes were often slow, discrete, and isolated.

A data correction might affect a single report. A system update would be scheduled, reviewed, and deployed infrequently. The consequences of change were usually visible and reversible.

AI changes this dynamic.

Small changes can propagate widely. A data update may influence multiple models. A configuration change may alter downstream decisions. A transaction performed at scale can retrain behavior, not just record it.

In this environment, governance cannot treat “making changes” as a low-risk activity.

The Maker role exists because change itself is now a form of decision-making.

## Makers Operate Within Guardian-Defined Constraints

Makers do not act in a vacuum.

Their authority is shaped by the decision classes and security profiles defined by Guardians. These constraints specify:

- what types of changes are permitted,
- which data may be modified,
- under what conditions automation may occur,
- and when escalation is required.

This relationship is intentional.

Guardians define ***where*** change is **allowed**.
Makers determine ***how*** change is **executed**.

When constraints are clear, Makers can move quickly. When boundaries are ambiguous, Makers slow down or take risks unknowingly. The GMS Framework resolves this tension by making constraints explicit and enforceable.

### Data Stewards as Makers

The Data Steward role fits naturally within the Maker state.

Data Stewards are responsible for:

- correcting errors,
- resolving inconsistencies,
- applying business rules,
- maintaining reference data,
- and ensuring data quality over time.

Under the GMS Framework, these activities are recognized as governance-relevant actions, not administrative tasks. Stewardship becomes visible, accountable, and connected to downstream outcomes.

This elevation matters.

When stewardship is treated as invisible labor, risk accumulates quietly. When it is treated as Maker activity, governance can account for its impact.

Transactional Roles Are Also Makers

Not all Makers think of themselves as working with data.

A customer service agent updating a record.
A claims processor approving an adjustment.
A logistics coordinator rerouting an order.

These actions are often framed as operational, not governance related. In AI-enabled systems, that distinction no longer holds.

Transactional actions:

- modify data,
- influence model behavior,
- and shape future decisions.

Under the GMS Framework, these roles are recognized as Makers when they change system state. This does not mean they carry the same authority as engineers or data stewards. It means their actions are governed appropriately based on decision class and security profile.

Recognition enables protection.

When transactional Makers are governed intentionally, they are supported by clear constraints rather than exposed to hidden risk.

## Automation and the Maker Role

Makers may be human or automated.

Scripts, workflows, and AI agents often perform Maker actions at scale – updating records, enriching data, retraining models, or executing predefined changes. These automated Makers operate under Guardian-defined constraints and inherit the same accountability structure.

Automation does not remove responsibility.

It changes how responsibility is exercised.

Human Makers remain accountable for:

- defining automation logic,
- validating its behavior,
- monitoring outcomes,
- and intervening when constraints are breached.

This preserves governance even as execution accelerates.

What a Maker Actually Does

The Maker role is responsible for changing the system within permitted bounds.

This includes:

- modifying data or metadata,
- adjusting logic, rules, or configurations,
- implementing Guardian-defined constraints,
- executing transactions that alter system state,
- and validating that changes behave as intended.

Makers do not define policy.
They do not set risk tolerance.

They translate intent into action.

## Makers Are Not Free Agents

It is important to state what the Maker role is not.

Makers do not have unlimited authority.
They do not decide which constraints apply.
They do not override Guardians by default.

Their freedom comes from clarity, not absence of control.

When Makers understand the boundaries, they move faster and make better decisions. When boundaries are unclear, risk increases for everyone.

# Maker Is a State of Responsibility

Like Guardian, Maker is a state, not a title.

A person may act as a Maker while updating data, then transition into a Seeker role when using outputs, or into a

Guardian role when defining constraints for future work. What matters is that responsibility is tied to activity, not identity.

Automated processes may also act as Makers, but accountability remains human.

This keeps governance intact without slowing execution.

Why Makers Enable Scale

AI systems scale change.

Without clear Maker accountability, scale amplifies risk. With it, scale amplifies value.

The Maker role ensures that change is:

- intentional,
- bounded,
- traceable,
- and reversible where possible.

This is how organizations evolve systems without losing control.

The Guardian defines the line.
The Maker moves the system forward within it.

Let's review The Maker Principle graphic again. (also see figure 2-7-1 for the first introduction)

The Maker is the key to governance at scale under controlled principles.

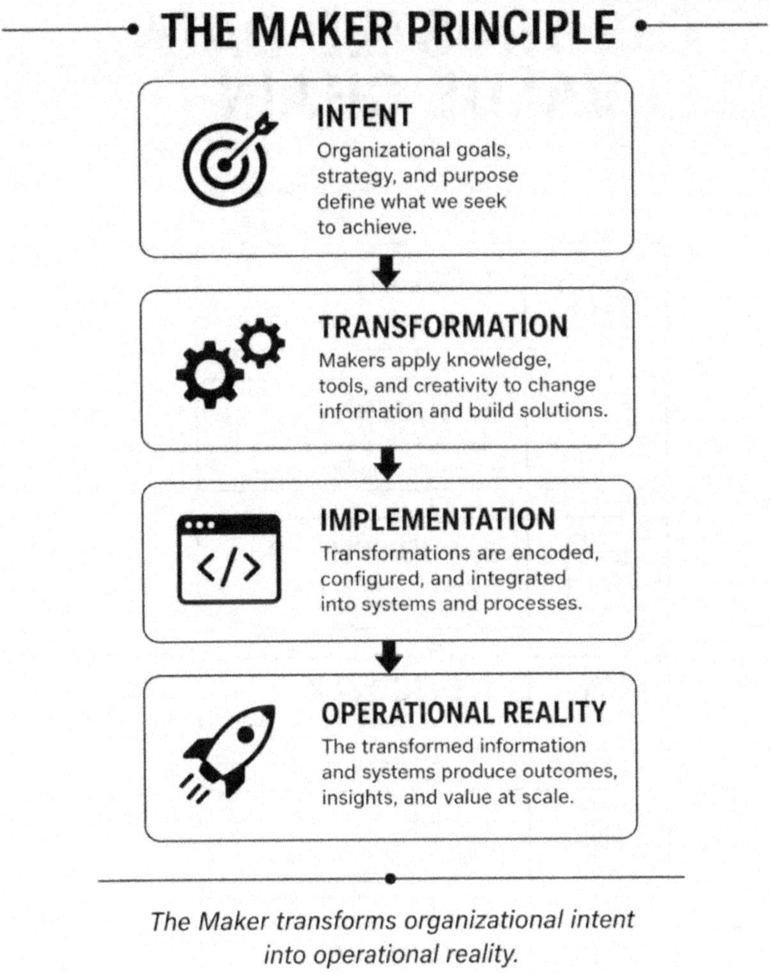


*Figure 5-19-1: The Maker Principle*

*Figure 5-20-1: Trusted Data Value Chain*

# Chapter 20: From Consumer to Seeker

Using Intelligence Without Abdicating Judgment

The Guardian defines boundaries and the Maker changes the system; the Seeker is where consequence appears.

The Seeker is the role most people recognize instinctively. It is the role of the decision-maker, the analyst, the operator, the executive, the clinician, the case worker, the manager. It is where outputs are consumed, insights are interpreted, and action is taken in the world.

What changes in AI-enabled environments is not that people stop being Seekers.

It is that seeking no longer happens in isolation from the system that produced the answer.

## From Consumer of Reports to Accountable Information Gathering

In legacy environments, Seekers consumed outputs that were largely static.

Reports were generated on a schedule. Dashboards refreshed periodically. Analyses were reviewed, discussed, and acted upon with time for deliberation. The distance between data production and decision-making allowed ambiguity to persist.

AI collapses that distance.

Outputs are generated continuously. Recommendations surface in real time. Decisions are often prompted, prioritized, or shaped by AI systems before a human consciously requests them.

In this environment, the Seeker role becomes an important part of the governance triad.

Seeking is no longer passive consumption.
It is an act that carries responsibility.

### What a Seeker Actually Does

The Seeker role is responsible for using outputs to make decisions and take action.

This includes:

- interpreting AI-assisted recommendations,
- selecting which outputs to rely on,
- applying judgment in context,
- deciding when outputs are sufficient and when they are not,
- and acting on those decisions in the real world.

Seekers do not build systems.
They do not define constraints.

They decide what to do with what is presented.

This is where impact occurs.

## Seekers and Permitted Inference

A defining feature of the Seeker role is operating within **permitted inference**.

Guardians define what conclusions may be drawn from certain data and under what conditions. Makers implement systems that enforce those constraints. Seekers are responsible for staying within them.

This matters because AI outputs often invite overreach.

Models can surface correlations that appear meaningful but are contextually inappropriate. Outputs can feel authoritative even when uncertainty is high. Seekers must recognize when an answer is fit for purpose and when it is not.

This is not a technical skill.

It is a governance responsibility.

**AI Assistance Does Not Replace Seeker Accountability**

AI systems may surface insights, rank options, or recommend actions.

They do not assume Seeker responsibility.

Even when AI agents initiate actions or trigger workflows, accountability for outcomes remains human. Seekers are responsible for understanding:

- what the output represents,
- what assumptions underpin it,
- what data was included or excluded,
- and what consequences may follow from acting on it.

This preserves judgment. And judgment relies on Critical Thinking abilities.

Automation may accelerate seeking, but it does not absolve it.

Seekers Are Not Endpoints

It is tempting to think of Seekers as the final step in a pipeline.

They are not.

Seeker actions feed back into the system. Decisions taken become data. Outcomes influence future behavior. This feedback loop is one of the defining features of AI-enabled environments.

Because of this, Seeker behavior must be governable.

Unchecked use creates drift. Overreliance creates fragility. Misinterpretation creates harm. The GMS Framework recognizes these risks and makes Seeker activity visible rather than assumed.

## Seeker Responsibility Under Pressure

The Seeker role is most tested under pressure.

When time is limited.
When stakes are high.
When outputs conflict with intuition.

In these moments, governance must support judgment rather than undermine it.

Seekers need:

- confidence in the boundaries Guardians defined,
- trust in the changes Makers implemented,
- and clarity about when escalation is required.

When these conditions are present, Seekers can act decisively without fear of hidden consequences.

What the Seeker Role Is Not

Seekers are not passive recipients of insight.
They are not absolved because a model suggested an action.
They are not expected to understand system internals.

They are expected to exercise judgment within defined bounds.

This distinction protects both individuals and organizations.

**Seeker Is a State of Responsibility**

Like Guardian and Maker, Seeker is a **state**, not a title.

A person may act as a Seeker when interpreting outputs, then move into a Maker role when correcting data, or into a Guardian role when redefining constraints based on observed outcomes.

AI agents may generate outputs, but that alone does not make them Seekers. Responsibility for decisions remains human, even when execution is automated.

This ensures that accountability does not disappear as systems become more capable.

Why the Seeker Role Completes the Framework

Without Guardians, there are no boundaries.
Without Makers, there is no controlled change.
Without Seekers, there is no responsible action.

The Seeker role ensures that intelligence is **used**, not just produced.

It anchors governance at the point where outcomes matter most. It reinforces that AI does not eliminate human responsibility - it concentrates it.

This is how organizations use intelligence without surrendering judgment.

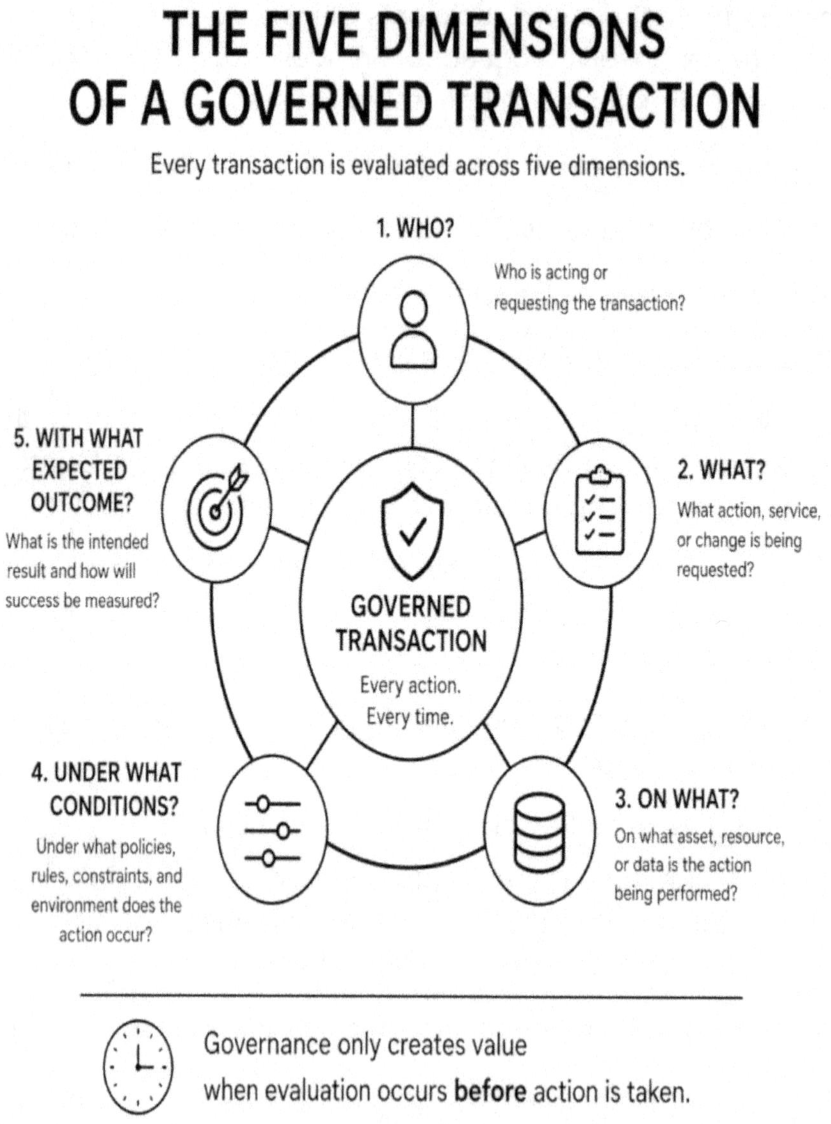


*Figure 5-21-1: The Five Dimensions of a Governed Transaction*

# Chapter 21: The GMS Interaction Model

Governing in the Moment When Intelligence Moves Faster Than Structure

Up to this point, two assertions have been made.

First, that AI collapses legacy governance models - partly because of speed, partly because of automation.

Second, that the Guardian–Maker–Seeker Framework enables a different kind of governance - one that operates in the moment rather than after the fact.

This chapter explains why both are true, and how the interaction between these roles makes continuous governance possible.

Why AI Breaks Sequential Governance

Legacy governance assumes sequence.

Work is performed.
Results are reviewed.
Decisions are approved.
Controls are applied.

This model depends on time and separation. It assumes that:

- decisions are discrete,
- changes are visible before consequences occur,
- and humans remain the primary execution layer.

AI violates all three assumptions.

AI systems operate continuously. Decisions are not queued for review - they are executed as conditions arise. Automation allows actions to propagate without human pause. Feedback loops mean that today's decision becomes tomorrow's training signal.

In this environment, **governance that waits its turn arrives too late**.

Sequential oversight collapses because the system does not wait for it.

Why Automation Changes Accountability, Not Just Speed

Automation is often discussed as a productivity gain.

In governance terms, it is an accountability challenge.

When actions are automated:

- intent is defined earlier,
- execution is delegated,
- and consequences may surface far from the original decision.

Without explicit role interaction, responsibility becomes diffuse. Teams debate whether accountability belongs with the person who defined the rule, the system that executed it, or the individual who relied on the outcome.

The GMS Framework resolves this ambiguity by **decoupling responsibility from execution and tying it to role state**.

This is the critical shift.

The Core Interaction Pattern

At its simplest, the GMS interaction model operates as a loop:

1. **Guardians define constraints**
2. **Makers act within those constraints**
3. **Seekers use outputs within permitted inference**
4. **Observed outcomes inform Guardian refinement**

This loop is continuous, not linear.

Each role is active at different moments, sometimes exercised by the same individual, sometimes by different actors. Governance emerges from how these roles interact, not from any single role acting alone.

## The Five Dimensions of a Governed Transaction

The Five Dimensions define the runtime evaluation performed for every governed transaction.

Every governed transaction is evaluated by answering five questions:

1. **Who** is acting? (Guardian, Maker, Seeker, AI Agent, Human, Service Account, etc.)
2. **What** are they attempting?
3. **On what** data, model, or digital asset?
4. **Under what** policies, context, and constraints?
5. **With what expected outcome** (allow, deny, escalate, monitor, audit...)

Every governed transaction is evaluated across all five dimensions before action is permitted.

**That becomes the operational heartbeat of GMS.**

Rather than introducing another governance framework, the Five Dimensions define the evaluation algorithm through which every governed interaction passes.

**WHO:** When looking at "Who" is acting or requesting the transaction we must evaluate the identity, role, activity, trust level, authorization, security profile and accountability structure.

**WHAT:** What action, service or change is being requested. At this state we Evaluate the type of action, operation, purpose and lifecycle Change.

**ON WHAT:** On what asset, resource or data is the action being performed? Evaluate the asset type, classification sensitivity, ownership, classification, security guardrails, technical and business metadata, and context.

**UNDER WHAT CONDITIONS**: Under what policies, rules, constraints, and environment does the action occur? These are the external rules that provide a governed practice. Evaluate applicable policies, business rules, regulations, risk, location, time and consent.

**WITH WHAT EXPECTED OUTCOME:** What is the intended result and how will success be measured? What proof was generated that the transaction was properly executed? Evaluate the expected result, success criteria, impact and monitoring requirements.

Governance only creates value when evaluation occurs before action is taken in real time. Each transaction receives an immediate status of:

- Allow = Proceed

- Escalate = Needs Review
- Deny = Block Action
- Monitor = Track Activity
- Audit = Record event, may include awareness flag

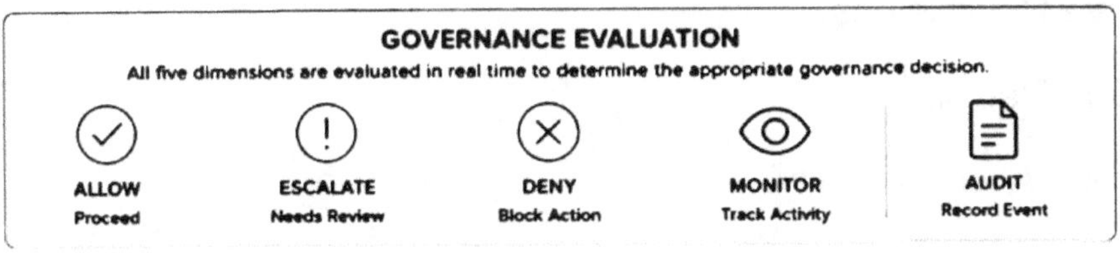


*Figure 5-21-2: Governance Evaluation*

**These five dimension must be present** in every transaction – every movement – every request – every decision. Each of these triggers is a moment of decision.

Every decision has context, intent, and consequence. It is critical that all five dimensions are applied to all data, without exception.

Governance is consistent when the same evaluation model is applied across all systems, users and AI agents. Risk must be managed at the right moment. Decisions are often timely or have regulatory guidance.

Decisions are evaluated before the action occurs, not after impact. Above all, accountability must be preserved. With the actors and AI personas flexibility to “switch hats” and play more than one role, it is crucial that every action is tied to the responsible actor and a measurable outcome.

Our goal is to turn as many of these important but repeatable tasks into automation. But for that to be feasible, automation must remain trustworthy. AI Agents and automated processes are governed by the same rules as humans.

And finally, Compliance becomes embedded in the code that moves data. Policies, rules, and regulations are enforced in real time, at the point of transaction.

Five Dimensions.

One Decision.

Total Accountability.

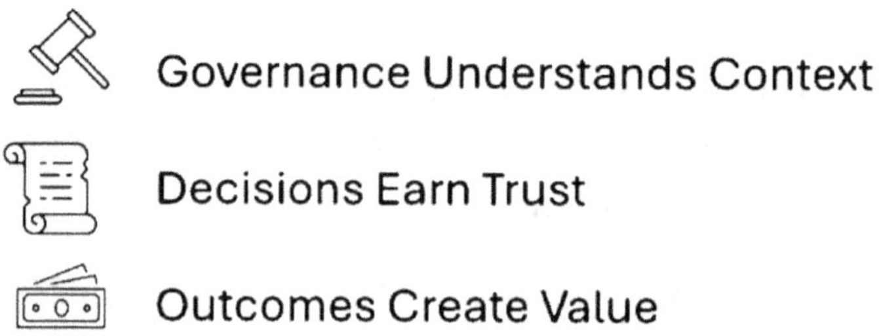


*Figure 5-21-3: Governance Creates Value*

The Five Dimensions define *what* must be evaluated. The next section explains *how* that evaluation occurs in real time.

How Governance Happens in the Moment

Every governed transaction begins with the Five Dimensions. The evaluation process transforms those dimensions into a governance decision that can be executed immediately. Governance occurs when **role boundaries are crossed**.

When a Maker attempts a change that exceeds permitted bounds, Guardian-defined constraints intervene. When a Seeker attempts to draw an inference outside permitted use, constraints surface context or require escalation. When outcomes deviate from intent, feedback loops inform Guardian reassessment.

This is not oversight layered on top of work.

It is **governance embedded into work**.

Because constraints are evaluated dynamically:

- decisions do not wait for committees,
- automation does not escape accountability,
- and humans remain responsible without being overwhelmed.

This is how governance keeps pace with AI.

## WHY THESE FIVE DIMENSIONS MATTER

A transaction is not just a request—
it is a moment of decision.

Every decision has context, intent,
and consequence. These five dimensions ensure that:

### Governance is consistent

The same evaluation model is applied across all systems, users, and AI agents.

### Risk is managed at the right moment

Decisions are evaluated before action, not after impact.

### Accountability is preserved

Every action is tied to a responsible actor and a measurable outcome.

### Automation remains trustworthy

AI agents and automated processes are governed by the same rules as humans.

### Compliance becomes embedded

Policies, rules, and regulations are enforced in real time, at the point of transaction.

---

**FIVE DIMENSIONS. ONE DECISION. TOTAL ACCOUNTABILITY.**

**Governance that understands context.**
**Decisions that earn trust.**
**Outcomes that create value.**

*Figure 5-21-4: Why the 5 Dimensions Matter*

## Why Role Interaction Matters More Than Role Definition

Defining roles alone is insufficient.

Governance emerges from **transitions between roles**.

A person may:

- act as a Guardian when defining decision classes,
- shift into a Maker role to implement logic,
- then move into a Seeker role when evaluating results.

Each transition carries different accountability.

The GMS Framework enforces clarity at these boundaries. It ensures that when responsibility shifts, so does authority and expectation. This prevents silent role blending, where individuals unknowingly carry conflicting responsibilities.

Automation respects the same boundaries.

AI agents may execute Maker actions or surface Seeker outputs, but they do not cross role boundaries independently. Accountability remains human, even when execution is delegated.

Why This Model Scales Where Others Fail

Legacy governance scales by adding process.

The GMS Framework scales by **making responsibility explicit**.

As systems grow more complex:

- constraints remain consistent,
- accountability remains traceable,
- and governance remains situational.

The model does not require centralization. It does not require constant review. It requires clarity.

This clarity allows organizations to move faster **because** governance is present, not despite it.

## From Collapse to Coherence

AI collapses prior governance models because they were designed for a slower world.

The Guardian–Maker–Seeker interaction model works because it reflects how work actually happens now:

- continuously,
- across humans and machines,
- under changing context,
- with immediate consequence.

Governance succeeds when it aligns with reality.

The GMS interaction model does exactly that.

What follows will build on this foundation - showing how this interaction model manifests in architecture, policy, and real organizational scenarios.

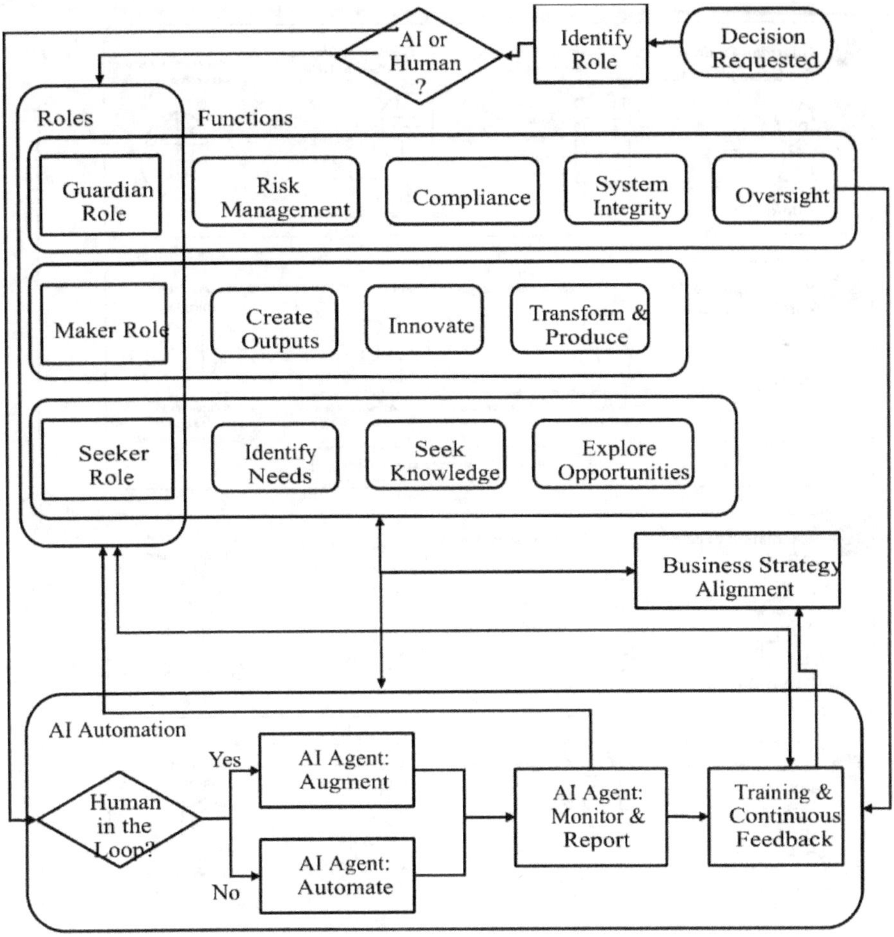


*Figure 5-21-5: GMS Interaction model sample*

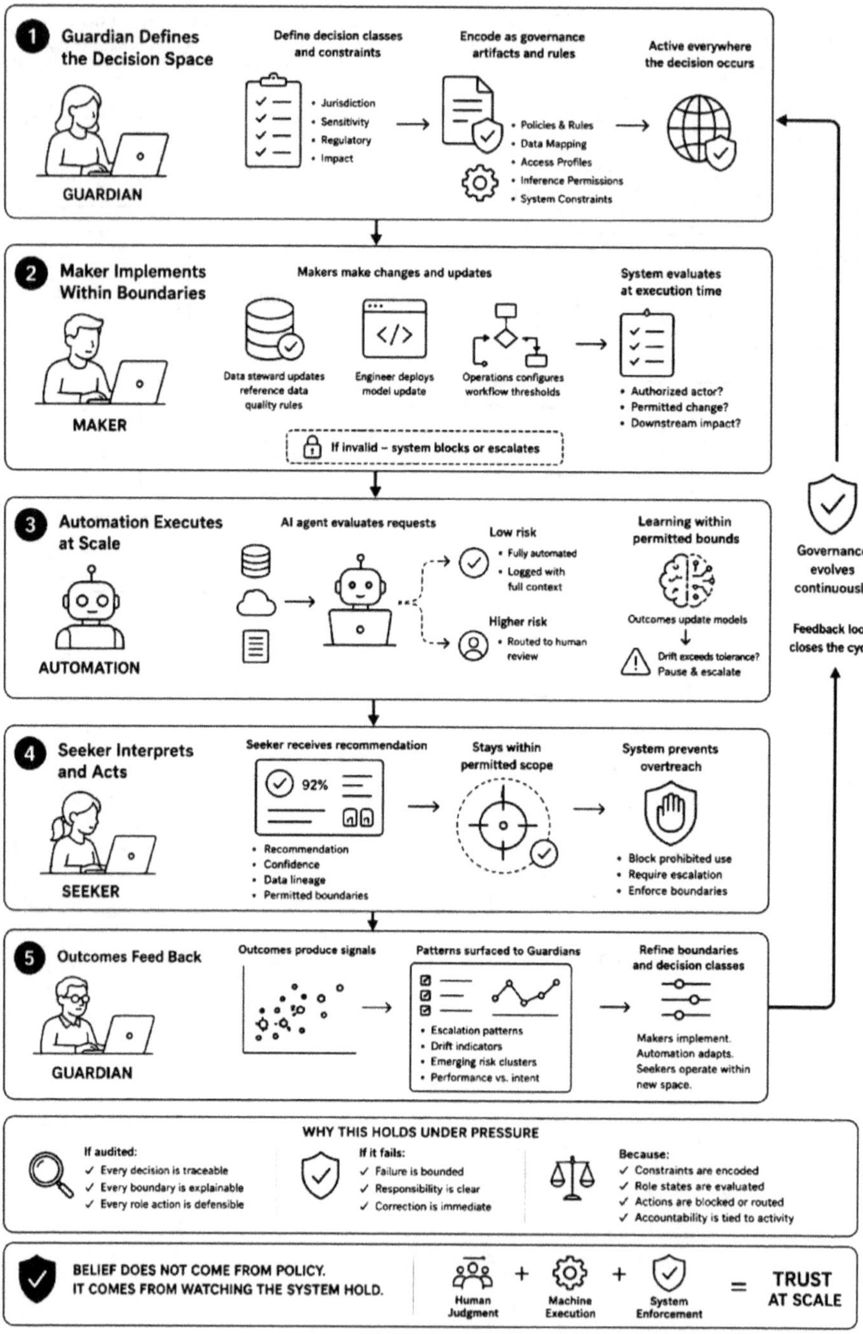
1 Guardian Defines the Decision Space
GUARDIAN
Define decision classes and constraints
• Jurisdiction
• Sensitivity
• Regulatory
• Impact
Encode as governance artifacts and rules
• Policies & Rules
• Data Mapping
• Access Profiles
• Inference Permissions
• System Constraints
Active everywhere the decision occurs
2 Maker Implements Within Boundaries
MAKER
Makers make changes and updates
Data steward updates reference data quality rules
Engineer deploys model update
Operations configures workflow thresholds
System evaluates at execution time
• Authorized actor?
• Permitted change?
• Downstream impact?
If invalid – system blocks or escalates
3 Automation Executes at Scale
AUTOMATION
AI agent evaluates requests
Low risk
• Fully automated
• Logged with full context
Higher risk
• Routed to human review
Learning within permitted bounds
Outcomes update models
Drift exceeds tolerance? Pause & escalate
4 Seeker Interprets and Acts
SEEKER
Seeker receives recommendation
92%
• Recommendation
• Confidence
• Data lineage
• Permitted boundaries
Stays within permitted scope
System prevents overtreach
• Block prohibited use
• Require escalation
• Enforce boundaries
5 Outcomes Feed Back
GUARDIAN
Outcomes produce signals
Patterns surfaced to Guardians
• Escalation patterns
• Drift indicators
• Emerging risk clusters
• Performance vs. intent
Refine boundaries and decision classes
Makers implement. Automation adapts. Seekers operate within new space.
Governance evolves continuously
Feedback loop closes the cycle
WHY THIS HOLDS UNDER PRESSURE
If audited:
✓ Every decision is traceable
✓ Every boundary is explainable
✓ Every role action is defensible
If it fails:
✓ Failure is bounded
✓ Responsibility is clear
✓ Correction is immediate
Because:
✓ Constraints are encoded
✓ Role states are evaluated
✓ Actions are blocked or routed
✓ Accountability is tied to activity
BELIEF DOES NOT COME FROM POLICY.
IT COMES FROM WATCHING THE SYSTEM HOLD.
Human Judgment
+
Machine Execution
+
System Enforcement
=
TRUST AT SCALE

*Figure 5-22-1: GMS in Motion*

# Chapter 22: The Belief Chapter

## An End-to-End GMS Scenario in Motion

Belief in governance does not come from diagrams.

It comes from watching a system hold under pressure.

Up to this point, the Guardian–Maker–Seeker Framework has been described as a pattern of responsibility, interaction, and accountability. This chapter shows how that pattern operates end to end – across people, systems, automation, and learning – without relying on goodwill or perfect behavior.

What follows is a single scenario. Nothing extraordinary happens. That is the point.

### The Scenario Context

A global services organization uses AI to support customer eligibility decisions across multiple jurisdictions. The decision impacts pricing, service access, and regulatory exposure. Some decisions are low risk. Others carry legal and reputational consequences.

The system includes:

- structured customer data,
- external reference data,
- AI models for risk scoring,

- automated workflows,
- and human review for defined cases.

Governance must function continuously.

## Step 1 – Guardian Defines the Decision Space

**(Business process + system constraint)**

Before any decision is made, Guardians act.

They define **decision classes** for eligibility determination, segmented by:

- jurisdiction,
- data sensitivity,
- regulatory exposure,
- and impact to the customer.

For each decision class, Guardians establish:

- permitted data sources,
- allowed model types,
- acceptable confidence thresholds,
- automation limits,
- and mandatory escalation triggers.

These are not documents alone.

They are translated into **governance artifacts:**

- policy-to-rule mappings,
- rule-to-data mapping,
- access control profiles,

- inference permissions,
- and system-enforced constraints.

This work is **manual in intent, but systemic in enforcement.**

Once defined, these constraints are active everywhere the decision can occur.

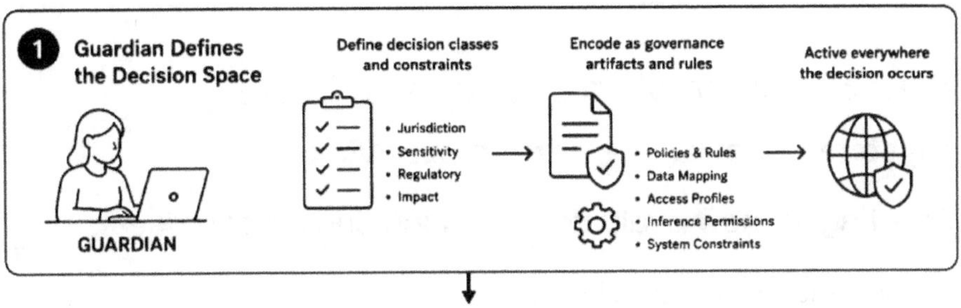


## Step 2 – Maker Implements Within Boundaries

**(System constraint + human execution)**

Makers now act.

A data steward updates reference data quality rules.
An engineer deploys a model update.
An operations team configures workflow thresholds.

Each Maker action is evaluated at execution time:

- Is the actor authorized for this change?
- Is the change permitted for this decision class?
- Does it affect downstream automated decisions?

If a proposed change violates Guardian-defined constraints, it cannot proceed without escalation. The system does not wait for review — it **prevents invalid state transitions.**

Makers move quickly because **boundaries are explicit.**

No one is guessing what is allowed.

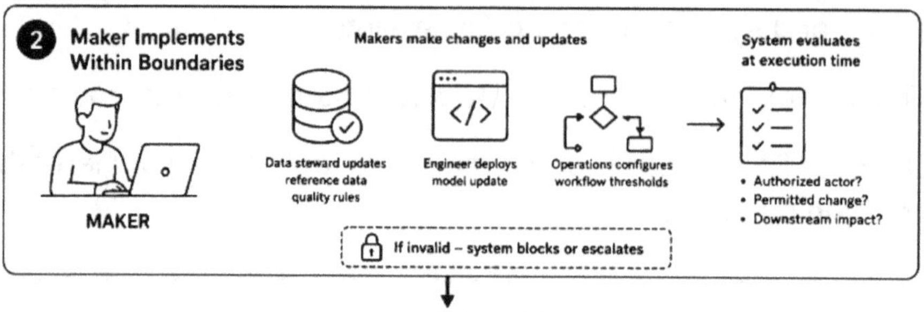


## Step 3 – Automation Executes at Scale

**(System constraint + learned behavior)**

An AI agent now evaluates incoming customer requests.

For low-risk decision classes:

- decisions are fully automated,
- execution proceeds without human review,
- actions are logged with full role context.

For higher-risk classes:

- automation operates only up to defined thresholds,
- confidence scores are evaluated,
- and cases exceeding limits are routed for human review.

The AI agent is not making governance decisions.

It is operating inside Guardian-defined constraints, implemented by Makers, and enforced by the system.

Learning occurs — models update based on outcomes — but **only within permitted bounds**. If drift exceeds tolerance, automation pauses and signals escalation.

This is enforcement through design.

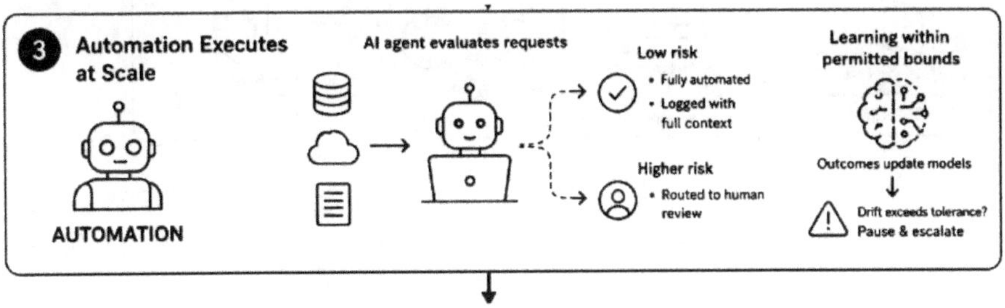


## Step 4 – Seeker Interprets and Acts

**(Business process + permitted inference)**

A human Seeker receives a recommendation.

They see:

- the recommendation,
- confidence indicators,
- data lineage,
- and permitted inference boundaries.

The system does not allow conclusions outside the approved scope.
The Seeker cannot unknowingly overreach.

If the Seeker attempts to:

- use the output for a prohibited purpose, or
- extend inference beyond what is allowed,

the system intervenes — either by blocking action or requiring escalation.

The Seeker's judgment is supported, not replaced.

Responsibility is clear.

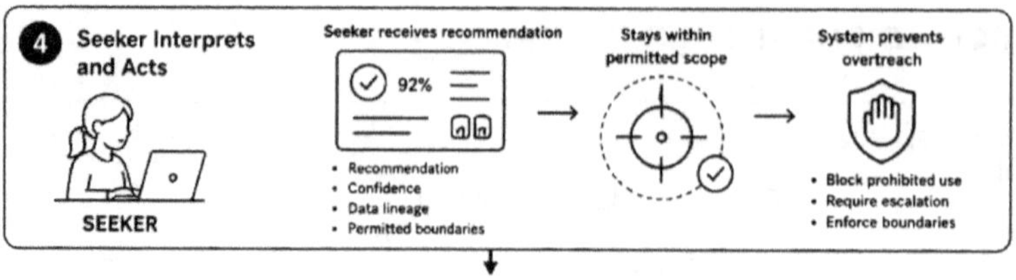


## Step 5 – Outcomes Feed Back

**(Learned behavior + Guardian oversight)**

Decisions produce outcomes.

Some reinforce expectations.
Others reveal edge cases.

These signals are aggregated and surfaced to Guardians:

- patterns of escalation,
- drift indicators,
- emerging risk clusters,
- and performance against intent.

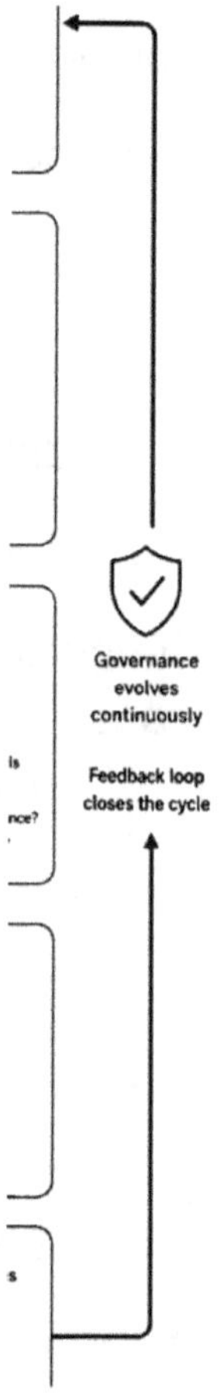


Guardians review **patterns**, not individual actions. This aligns with the committee oversight timelines and actions. Patterns and feedback consistently reviewed on a schedule to avoid escalations.

When boundaries need adjustment, Guardians refine decision classes or constraints. Makers implement the updates. Automation adapts. Seekers operate within the new space.

Governance evolves without disruption.

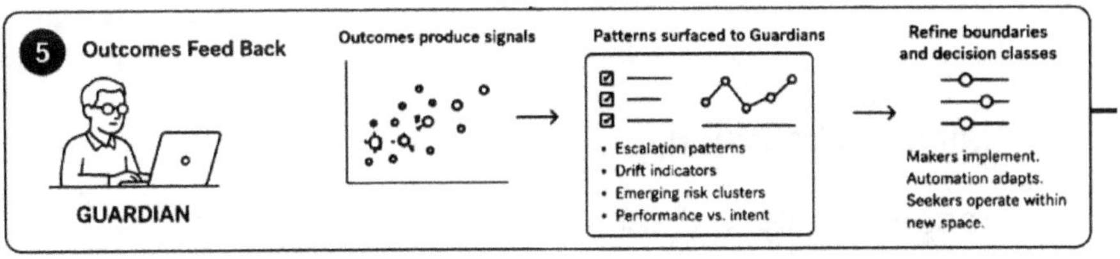


## Where Enforcement Actually Occurs

At no point does enforcement depend on:

- trust alone,
- training alone,
- or retrospective review alone.

Enforcement occurs because:

- **constraints are encoded**,
- **role states are evaluated dynamically**,
- **actions are blocked or routed by design**, and
- **accountability is tied to activity**.

This is not surveillance.
It is structure.

## Why This Holds Under Pressure

If this scenario is audited:

1. every decision is traceable,
2. every boundary is explainable,

3. every role action is defensible.

If it fails:

- failure is bounded,
- responsibility is clear,
- correction is immediate.

This is why the framework works.

Not because people behave perfectly, but because the system **does not allow irresponsible behavior to propagate**.

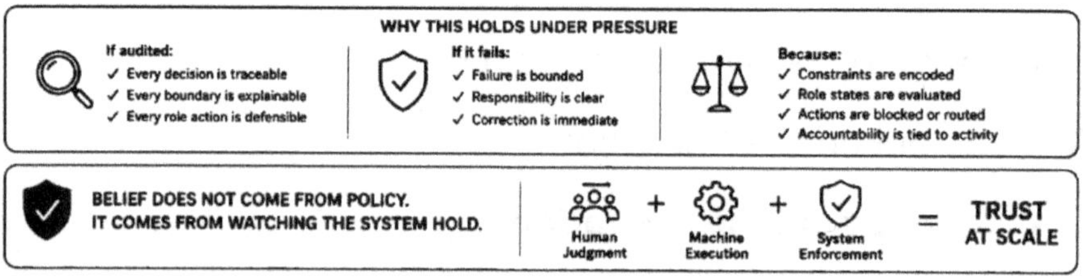


## Belief Earned

If governance required constant human vigilance, it would fail.

If automation were unconstrained, it would fail.

The Guardian–Maker–Seeker interaction works because it aligns:

- human judgment,
- machine execution,
- and system-level enforcement.

Governance becomes continuous, contextual, and real.

Belief does not come from policy.

It comes from watching the system hold.

# Chapter 23: Case Studies

The following three examples examine the framework from three different viewpoints: operations, architecture, and assurance.

## Case Study 1: Operational Example – When the System Fails | Governance Under Stress

Belief in governance is tested when something goes wrong.

Not when models behave as expected.
Not when automation stays within bounds.
But when outcomes surprise, assumptions fail, or pressure arrives from outside the organization.

This is where legacy governance collapses most visibly.

**The Failure Moment**

An AI-driven eligibility decision is challenged by a regulator.

The question is not whether the decision was correct.
It is whether it was **defensible**.

What data was used?
Who approved the automation?
Why was human review bypassed?
What safeguards were in place?

In legacy systems, these questions trigger reconstruction. Emails are searched. Committees are convened. Narratives are assembled after the fact.

In the GMS interaction model, the answers already exist.

**How Failure Is Contained**

When a failure occurs, the system does not search for blame.

It surfaces **role context**.

- Which Guardian-defined decision class was active
- Which constraints were enforced
- Which Maker changes were in effect
- Which Seeker action triggered the outcome

This context is not inferred. It is recorded as part of normal operation.

The failure is bounded because constraints limited scope. The response is targeted because accountability is explicit. The correction is fast because Guardians adjust boundaries rather than rewriting policy.

Failure becomes a governance signal, not a governance crisis.

This is how systems survive pressure without freezing.

## Case Study 2: Architectural Example – From Interaction to Architecture | How GMS Shows Up in Real Systems

The GMS Framework does not require a single architecture.

It requires **specific architectural capabilities**.

This chapter translates the interaction model into components leaders and architects already understand.

**What Must Exist in the Architecture**

Regardless of platform or vendor, systems must support:

- **Role-state awareness**
  Systems must know whether an action is Guardian, Maker, or Seeker activity.
- **Decision class tagging**
  Actions must be evaluated in context – not all decisions are equal.
- **Constraint enforcement points**
  Governance rules must block, route, or require escalation at execution time.
- **Traceability by default**
  Context must be captured as a byproduct of work, not a forensic exercise.
- **Feedback loops**
  Outcomes must inform boundary refinement, not just performance metrics.

These are not exotic requirements.

They already exist in pieces across IAM systems, workflow engines, data platforms, and MLOps pipelines. The GMS Framework **aligns them** to governance intent.

This is why it is an architectural pattern, not a product.

## Case Study 3: Audit Example – Defensibility Under Scrutiny | Standing Up to Audit, Regulation, and Reality

Governance ultimately proves compliance.

Regulators do not ask whether a framework exists. They ask whether it works.

The GMS Framework is designed for **epistemic defensibility**.

Epistemic defensibility means that decisions can be justified under scrutiny using evidence, context, and clearly assigned responsibility - not reconstructed after the fact through narrative or interpretation.

Key components of this concept include:

- **Justification:** The belief is backed by valid arguments, evidence, or reliable methods of inquiry.
- **Reasonableness:** The viewpoint is rational and coherent, holding up under critical examination.
- **Support:** There are sufficient reasons for holding the belief, moving it beyond mere opinion.
- **Resistance to Counterarguments:** The position can be defended against objections.

In essence, a position is "defensible" in this sense if it is worthy of acceptance based on the quality of the information or reasoning behind it.

**What Defensibility Looks Like**

Under scrutiny, the organization can show:

- why automation was permitted,
- under which constraints it operated,
- where human judgment intervened,
- and how responsibility was assigned.

This is not storytelling. It is evidence.

Because the framework governs *activity*, not intent, it produces defensibility as a side effect. Decisions are not justified after the fact — they are contextualized at the moment they occur.

This is the difference between compliance theater and operational governance.

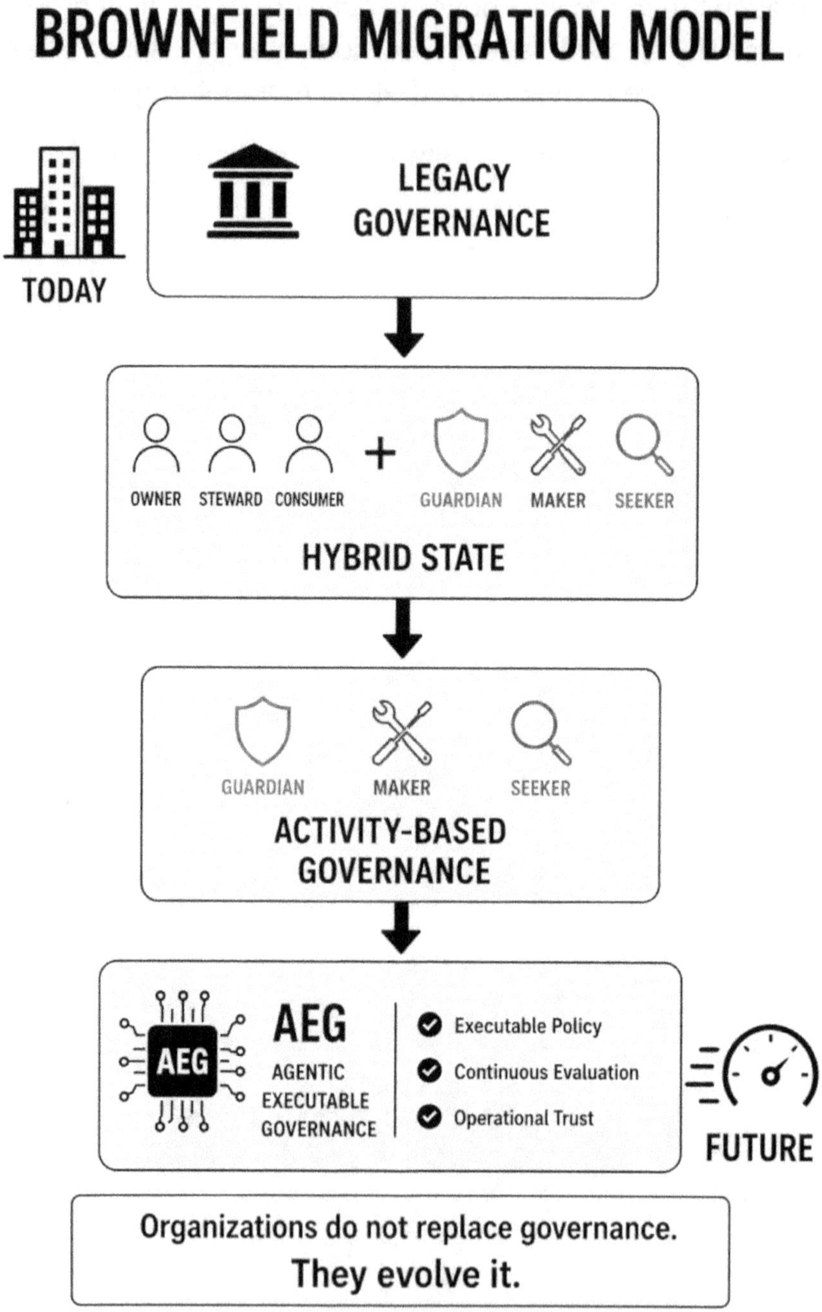


*Figure 5-24-1: The Brownfield Solution*

# Chapter 24: The Brownfield Problem - Governing What Already Exists

You must be asking the Critical Question: “What About Everything That Already Exists?”

This is where many governance frameworks fail spectacularly.

They assume a reset – reworking systems, assigning new roles, writing more policies and processes in every system where data exist.

The EZ-AI-Ready GMS Framework explicitly does **not** require a reset.

**GMS Is Overlay Governance, Not Rip-and-Replace**

The framework was designed for **brownfield reality**.

Most organizations already have:

- legacy data platforms,
- existing models,
- manual processes,
- embedded automation,
- and years of accumulated decisions.

The GMS Framework does not require rewriting them.

It requires **making existing decisions visible, classifying them, and gradually placing them under governed execution.**

## How Existing Assets Are Brought In

Existing systems are incorporated by:

1. **Classifying existing decisions**
   - What decisions already exist?
   - Which are automated?
   - Which are human?
   - Which carry risk?
2. **Assigning Guardian intent retroactively**
   - Decision classes are defined for what already happened
   - Constraints are applied prospectively, not rewritten historically
3. **Instrumenting enforcement points**
   - Controls are added where action occurs
   - Not everywhere at once
   - Not uniformly
4. **Gradual role visibility**
   - Makers and Seekers are recognized, not reorganized
   - Accountability becomes explicit without disrupting operations

This is why the framework can be adopted incrementally.

The fact is that most organizations have not documented these rules. Sometimes they haven't documented processes. This is a perfect case for AI to help by evaluating origination to outcome:

- A decision framework can be built.
- Process documentation can be inferred.
- Rules and policies can be aligned to the central decision and governance layer.

These must still be reviewed, but that is a much easier lift than to create these items from scratch. The process inference capability operates as Maker functions under Guardian-defined constraints, with accountability remaining human even when discovery is automated.

## Why Incremental Transition Works

Because governance does not require perfection.

It requires **clarity at the margin**.

Once:

- boundaries are explicit,
- enforcement exists at key points,
- and accountability follows activity,
- legacy systems become governable without being rebuilt.

The framework grows outward from the highest-risk decisions first.

That is how successful organizations change.

## Closing the Belief Loop

The GMS Framework works:

- not because it imagines a clean slate,

- not because it demands ideal behavior,
- but because it accepts reality and governs it anyway.

It governs what already exists.
It shapes what comes next.
And it holds when pressure arrives.

That is the test.

# Chapter 25: From Governance Theory to Structural Reality

At this point, it's time to stop and take inventory. What have we examined, what have we learned, and what questions have we answered? We have not avoided the hard questions; we have confronted them directly.

We have examined:

- The collapse of paperwork governance
- The acceleration introduced by AI
- The mismatch between static roles and dynamic decisions
- The necessity of in-the-moment accountability
- The interaction of Guardian, Maker, and Seeker
- The enforcement question – how governance holds under pressure
- The brownfield reality – governing what already exists

None of these are theoretical problems.

They are operational realities.

The central claim of Part V has been simple:

AI collapses legacy governance not because governance is unnecessary – but because governance was structured for a slower world.

Committees assumed time.
Policy assumed review cycles.
Architecture assumed separation between decision and consequence.

AI removes that separation.

Decisions now occur:

- Continuously
- Across humans and machines
- At scale
- Under regulatory and reputational exposure

The response cannot be more paperwork. It must be structural.

## What We Have Established

The Guardian–Maker–Seeker Framework is not:

- A software product
- A consulting methodology
- A role rebranding exercise
- A theoretical governance model

It is a pattern that aligns authority with activity.

Guardians define decision classes and constraints.
Makers implement and operate within those constraints.
Seekers act within governed boundaries.

Enforcement occurs at the moment of action – not after failure.
Governance becomes situational rather than positional.

We have shown:

- How the framework operates in motion
- How it holds under regulatory scrutiny
- How it overlays brownfield systems
- How incremental adoption works
- How AI itself can assist in surfacing undocumented decisions

We have answered the questions about what is needed and how this framework fulfills that need. We have demonstrated that the framework sets a pattern for governance.

It enforces because constraints are encoded.
Because role states are evaluated dynamically.
Because activity is governed — not intention.

But this still leaves something unfinished. Perhaps a more concrete answer to the question, “But how does this system enforce anything?”

## The Structural Question Beneath Governance

If governance must operate continuously,
and if decisions must be bounded at execution time,
and if meaning must be preserved across systems,

Then, something deeper must exist.

We have described the roles.
We have described the interaction.
We have described the enforcement.

But what is the structural unit that makes this possible?

What is the atomic thing being governed?

What must be standardized before governance can scale?

Without answering that, governance remains procedural.

With that answer, governance becomes infrastructure.

## The Bridge to Architecture

Up to now, we have operated at the level of:

- Responsibility
- Authority
- Decision classification
- Enforcement logic

In the next section, we move one layer deeper.

We move from governing decisions to governing meaning itself. Because AI does not break governance merely through speed. It breaks governance through semantic fragmentation. We see multiple identifiers for the same object. inconsistent definitions across systems, ambiguous data reused beyond original intent. Every social contract depends on shared meaning.

If meaning is unstable, governance cannot scale.

If meaning is standardized, governance becomes structural.

The next section introduces the deeper pattern beneath all of this. It is not new. It has appeared before – in trade, in scholarship, in data, and in language.

## Why This Matters Now

AI does not demand new ethics. It demands structural clarity. The Guardian–Maker–Seeker Framework provides behavioral and governance alignment. But behavioral alignment must sit on top of standardized meaning.

Otherwise:

- Guardians cannot define consistently.
- Makers cannot encode reliably.
- Seekers cannot reason defensibly.

Governance cannot float. It must anchor to something atomic. Part VI reveals that anchor.

And once seen, it becomes impossible to unsee.

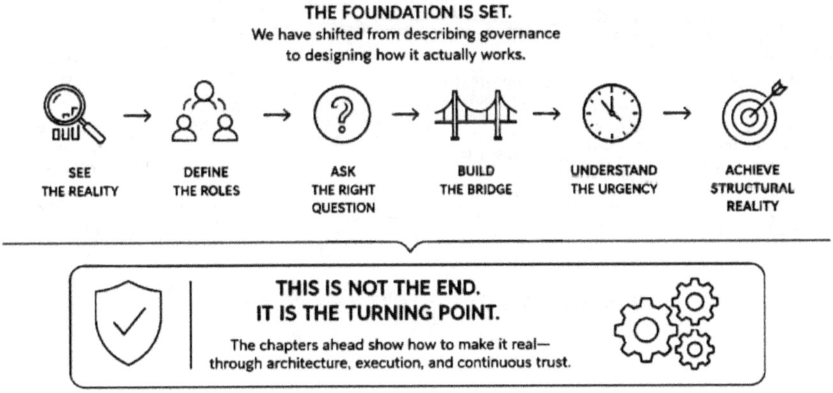


*Figure 5-25-1: The Turning Point*

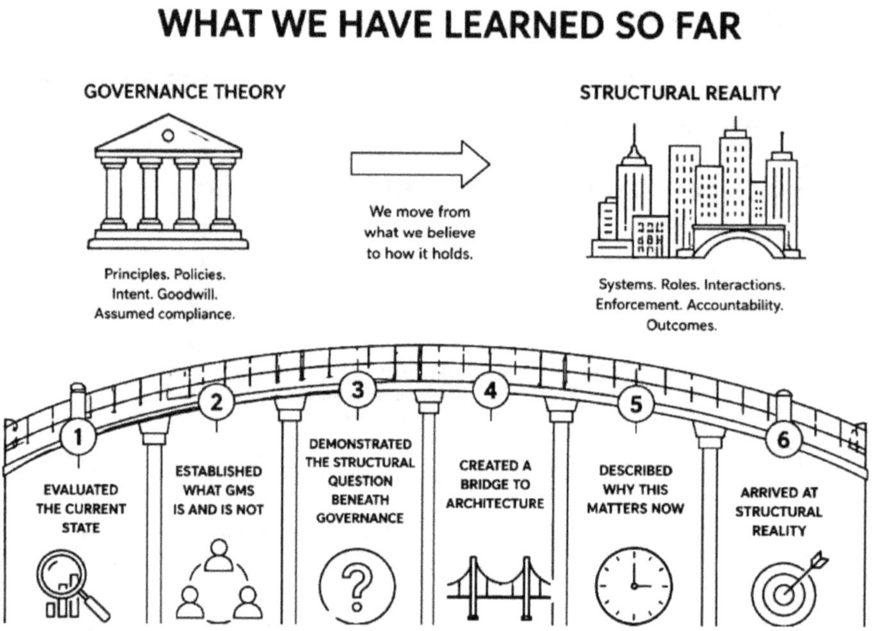


*Figure 5-25-2: From Theory to Reality*

# Part VI: The Unit Scaled the World

Civilization does not advance because systems become more complex.
It advances because complexity is built upon standardized units.

The industrial world learned this lesson in steel.
The intellectual world learned it in paper.
The data world is learning it in triples.
The linguistic world learned it in the word.
The AI era will only scale when meaning is standardized

We have already seen this pattern. Now we make it explicit.

- Malcom McLean did not invent trade. He standardized the container.
- Niklas Luhmann did not invent notetaking. He standardized the unit of thought.
- The RDF working group did not invent data. They standardized the fact.

- The Oxford English Dictionary did not invent language. It standardized the word.

In each case, the breakthrough was not invention.
It was universalization.

A simple unit became atomic.
Atomic became standardized.
Standardization forced infrastructure to rebuild.
Rebuilt infrastructure enabled emergent scale.

The pattern repeats across domains.

When we recognize it, it feels obvious.
When we ignore it, systems fragment.

These patterns are not academic. They are the structural precedent for semantic containers – and the reason governance can operate at the point of transaction.

This chapter reveals that pattern deliberately.

*Acknowledgement to Tony Seale, "The Knowledge Graph Guy," for the framing that inspired this synthesis. Extending the pattern across additional domains provides a more complete view of its universality.*

# Chapter 26: Standardizing Units

## The Shipping Container

Before containerization, global trade already existed. But it was slow, manual, and fragile. Cargo was loaded bale by bale, crate by crate. Every item required its own handling method. Every transfer required interpretation, coordination, and labor.

Ports were not just transfer points. They were points of friction.

Malcom McLean was a truck driver delivering goods to shipyards.
He spent hours waiting his turn as cargo was unloaded piece by piece. What he saw was not inefficiency in labor. He saw inconsistency in structure.

Each shipment was different.
Each required a different method.
Nothing was standardized.

He asked a simple question:

**What if the cargo did not matter?**

**What if everything moved in the same way?**

That question reframed the problem.

McLean did not start with ships. He started with the unit.

A box.

A fixed size.
A consistent shape.
A universal interface for lifting, stacking, and transport.

If cargo could be placed inside that box,
then the system no longer needed to adapt to the cargo. The

cargo adapted to the system. From that insight, everything else followed.

- Trucks had to be redesigned to carry standardized containers.
- Ships had to be rebuilt to stack them efficiently.
- Ports required cranes capable of lifting entire units instead of individual goods.
- Rail systems, tunnels, and overpasses had to accommodate consistent dimensions.

This was not a local optimization. It was a systemic transformation. And it only worked because it was universal.

The International Organization for Standardization (ISO) formalized this through ISO 668, defining container dimensions and ensuring compatibility across ships, trucks, and trains worldwide.

The shipping container did not introduce cargo. It introduced a standardized unit of cargo. A consistent size. A universal locking mechanism. Interoperability across every mode of transport.

The container forced ports to redesign – but only once. It forced ships to change – but made them more efficient. It forced logistics systems to synchronize – but made them predictable.

Global trade did not expand because steel boxes were clever.

It expanded because the world rebuilt around a standard.

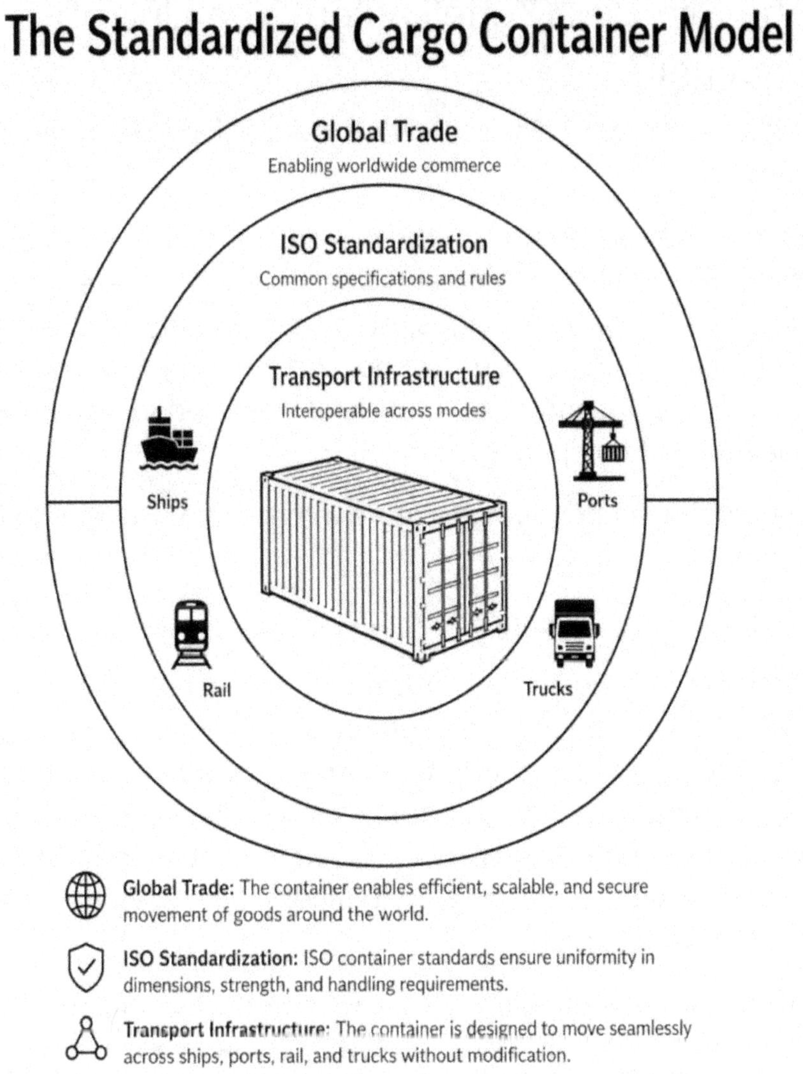


*Figure 6-26-1: The Standardized Cargo Container Model*

## The Slip-Box (Zettelkasten) The Unit of Thought

Niklas Luhmann did not invent notetaking. He restructured thinking.

Before the Zettelkasten, ideas were recorded in notebooks, journals, and files. Information was stored, but it rarely evolved. Notes were bound to pages, organized linearly, and difficult to reorganize once written. Connecting ideas across documents required effort, memory, and time. Knowledge accumulated, but it did not compound.

Luhmann was not trying to store information. He was trying to think with it.

That distinction changed everything.

Instead of writing pages, he wrote notes—small, self-contained slips of paper. Each note captured a single idea. Not a topic or a summary, but a discrete thought that could stand on its own. Every note was assigned a unique identifier, giving it a permanent address within the system. And each note could reference others, forming connections that extended beyond any single document.

Over time, this simple shift transformed a collection of notes into something far more powerful. It became a network of thought.

Luhmann's Zettelkasten grew to approximately 90,000 individual notes. He credited the system with enabling his extraordinary output—dozens of books and hundreds of articles. But the scale of output was not the breakthrough. The structure was.

Each note was atomic, addressable, and linkable.

The system did not rely on hierarchy or predefined organization. It grew through connection. Ideas were no longer trapped in documents; they were free to combine, evolve, and reappear in new contexts. A single note could connect to multiple lines of thinking. Previously unrelated ideas could intersect and produce something new.

The Zettelkasten did not improve notetaking. It standardized the unit of thought.

Once ideas were reduced to atomic units with stable identities and explicit relationships, thinking itself changed. Writing was no longer the act of starting from scratch, but the act of assembling and extending existing ideas. Thought was no longer linear; it became networked.

The slip-box became a system that could store ideas, retrieve them, recombine them, and generate new ones. What began as a method of organizing notes became a partner in thinking.

The Zettelkasten did not scale because paper was efficient. It scaled because ideas became atomic and linkable.

Just like the shipping container, the breakthrough was not the medium. It was the unit.

**Luhman did not create a filing system.**

**He created an operating system for ideas.**

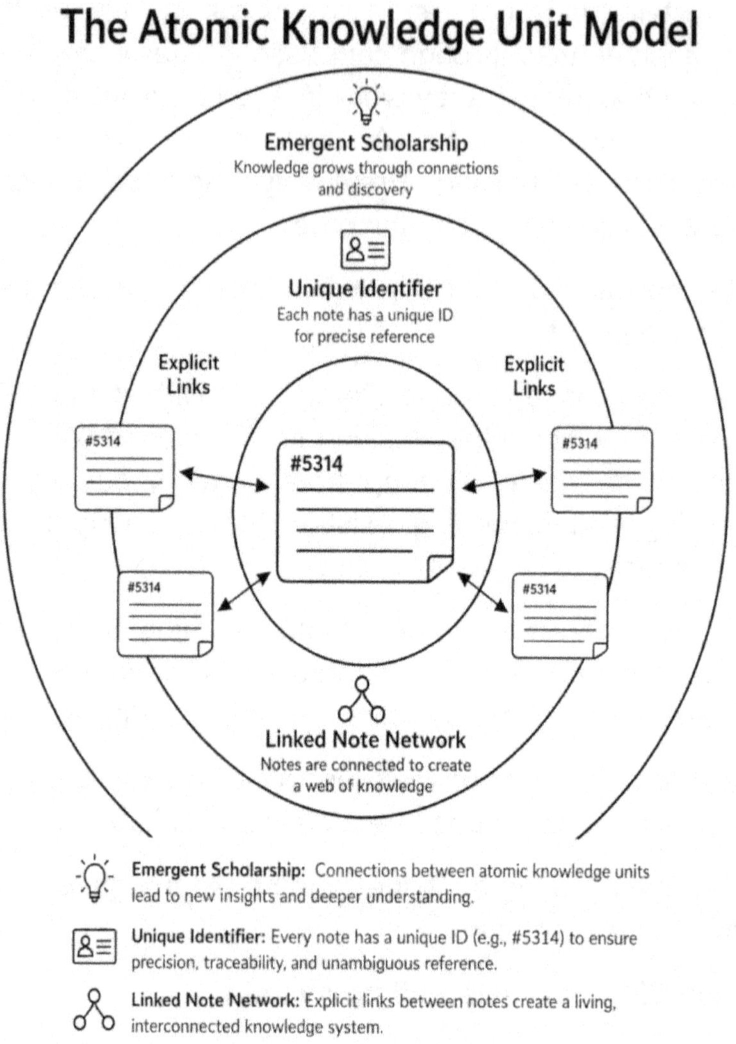


*Figure 6-26-2: The Atomic Knowledge Unit Model*

## The RDF Triple – The Unit of Atomic Fact

The RDF working group did not invent data.
They standardized the fact.

Before the Resource Description Framework (RDF) defined by the World Wide Web Consortium (W3C), data already existed everywhere – stored in databases, spreadsheets, applications, and documents. But it was fragmented, shaped by the systems that contained it. Each system defined its own structure, its own identifiers, and its own meaning.

Integration required translation.
Translation required interpretation.
And interpretation introduced inconsistency.

Data could be stored.
It could be queried.
But it could not easily be understood across systems.

The problem was not access.
It was meaning.

RDF approached this problem by changing the unit.

An RDF triple appears almost embarrassingly small:

Subject → Predicate → Object

"The Great Pyramid" → "is located in" → "Egypt."

At first glance, it does not seem powerful. It does not optimize a database. It does not summarize a dataset. It does not reduce complexity.

It standardizes a fact.

Each element receives a permanent identifier.

Shared ontologies define what those identifiers mean.

Graphs connect them into knowledge.

Each triple expresses a single, atomic assertion. A subject, a relationship, and an object – nothing more, nothing less. But when those elements are assigned universal identifiers and defined through shared ontologies, something important happens.

Meaning becomes portable. Each element is identified by a Uniform Resource Identifier (URI), ensuring global referenceability.

Facts are no longer bound to the system that created them. They can be referenced, linked, and understood across contexts. A triple created in one system can be interpreted in another – not because the systems are the same, but because the meaning is.

With identifiers, ontologies, and graph traversal, triples transform records into knowledge.

This is where the pattern reappears.

Like the slip-box, each triple is an individual unit. A single fact, self-contained, but connected. Each must be uniquely identified. Each can link to others. Over time, they form a network.

Not a collection of records – but a graph of meaning.

The triple becomes the card.
The graph becomes the slip-box.

And just as in Luhmann's system, the power is not in the individual unit, but in the connections between them.

Triples alone do nothing.

But triples that are standardized, linked, and governed create something fundamentally different. They allow meaning to be shared across models, systems, and organizations. They enable data to be recombined, traversed, and reasoned over.

They do not make data bigger.
They make it understandable.

The RDF triple does not scale because it is complex. It scales because it is simple.

Just like the shipping container.
Just like the slip-box.

The breakthrough was not the system.
It was the unit.

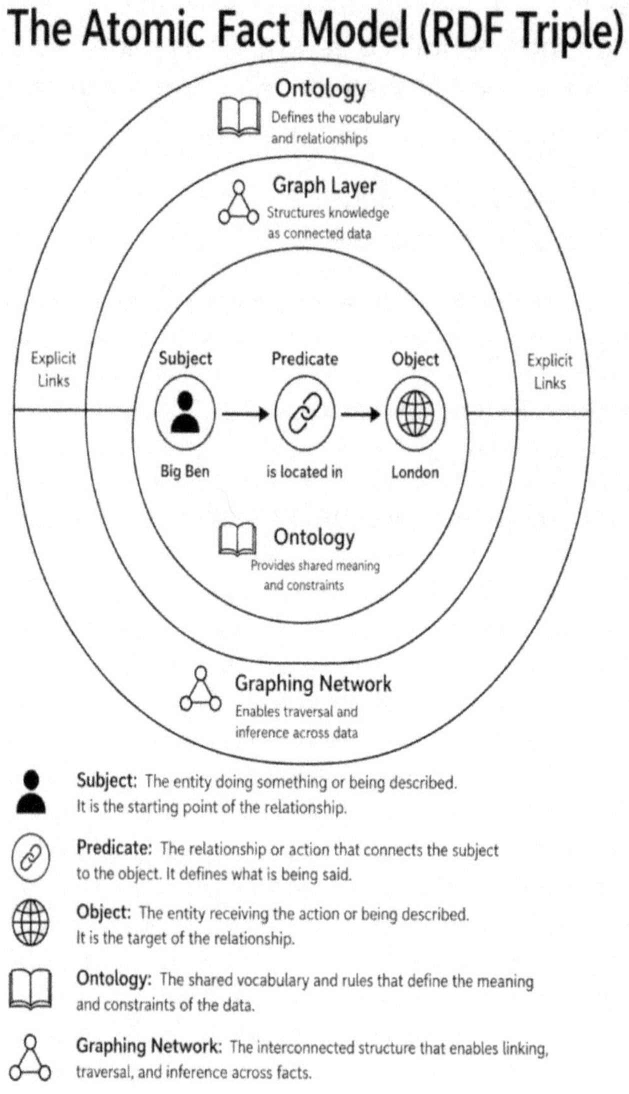


*Figure 6-26-3: The Atomic Fact Model – RDF Triple*

# The Standardized Word Model

## The Oxford English Dictionary - The Unit of Meaning

The Oxford English Dictionary did not invent language. It standardized the word.

For centuries, humans have faced the same problem we face in data today:
How do we ensure that meaning is shared?

Words evolve. Context shifts. Usage varies across regions, professions, and time. Without a common reference, communication fragments. Misunderstanding becomes inevitable.

The Oxford English Dictionary (OED) emerged as a response to that problem.

It did not attempt to simplify language.
It attempted to **anchor it**.

Each word in the OED is defined, contextualized, and traced through time. Meanings are not assigned casually; they are documented, evidenced, and refined. Usage examples ground definitions in reality. Etymology connects present meaning to historical origin.

The result is not just a dictionary.

It is a system of semantic authority.

When ambiguity arises, we do not invent new meaning. We consult a shared reference.

The OED provides a consistent interpretation of words across contexts. It allows writers, scholars, lawyers, and institutions to align on meaning – even when language itself is evolving.

In this way, it performs a critical function:

It standardizes the unit of meaning.

A word is no longer just a symbol.
It becomes a defined, referenceable concept.

But the OED was designed for humans.

It is not open.
It is not machine-computable.
It is not structured as a knowledge graph.

And yet, it works.

Human society has operated for centuries with a shared semantic foundation – not because meaning is simple, but because it is **standardized and referenceable**.

This is where the pattern becomes unavoidable.

There are modern attempts to create this same foundation for machines.

Projects like Wikidata assign identifiers and relationships at global scale.
WordNet organizes language into structured semantic relationships.
Schema.org provides shared vocabularies for the web.

Each contributes something important.

But none of them function as a universal semantic authority.

There is no Oxford English Dictionary for machines.

We have open ontologies.
We have knowledge graphs.
We have vocabularies.

But we do not have a universally accepted, authoritative standard for meaning.

The gap is not technical.
It is structural.

The table below illustrates the difference.

**Humans have the OED; Machines have fragments**

| Capability | OED | Wikidata | WordNet | schema.org |
|---|---|---|---|---|
| Authoritative meaning | ☑ | ⚠ | ⚠ | ✕ |
| Open | ✕ | ☑ | ☑ | ☑ |
| Machine-readable | ✕ | ☑ | ☑ | ☑ |
| Universal adoption | ⚠ | ⚠ | ✕ | ☑ |
| True semantic standard (human) | ☑ | ✕ | ✕ | ✕ |

*Figure 6-26-4: Distribution of available capabilities*

We have already solved the problem of meaning for humans. We have not yet solved it for machines.

Humans already solved semantic interoperability. Machines have not.

And until we do, AI will not scale through intelligence alone. It will scale through meaning.

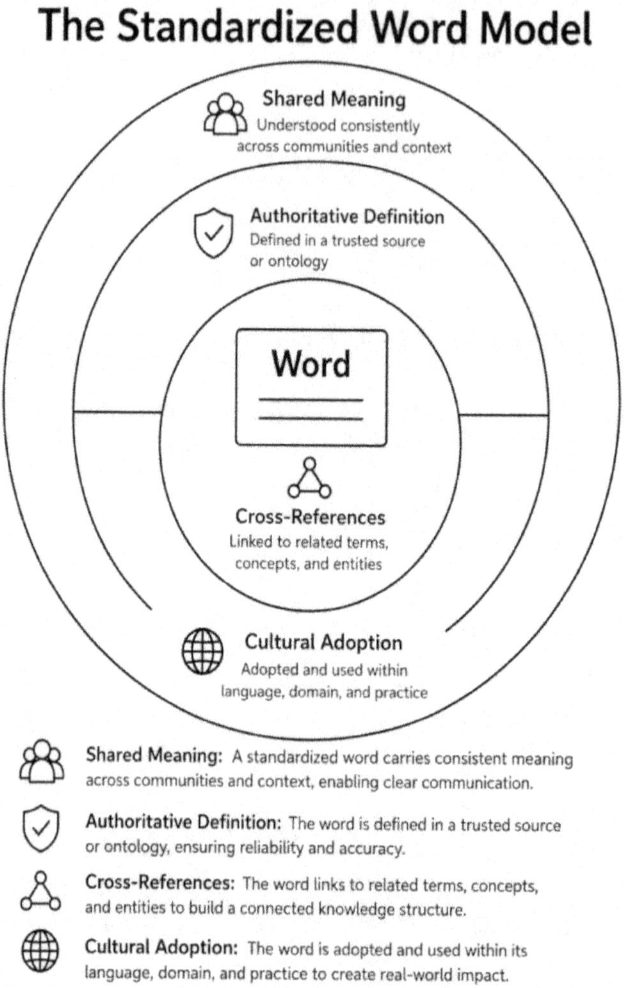


*Figure 6-26-5: The Standardized Word Model*

# Chapter 27: The Pattern Revealed

By now, the pattern should feel familiar.

We have seen it in steel.
We have seen it in paper.
We have seen it in data.
We have seen it in language.

Each domain solved a different problem.
Each arrived at the same answer.

A simple unit became standardized.
Standardization forced systems to rebuild.
Rebuilt systems enabled scale.

The shipping container did not expand trade because it was innovative.
It expanded trade because it was universal.

The Zettelkasten did not expand knowledge because it was organized.
It expanded knowledge because ideas became atomic and linkable.

The RDF triple did not expand data because it was structured.
It expanded data because facts became portable and composable.

The Oxford English Dictionary did not expand language because it defined words.
It expanded language because meaning became shared.

These are not separate innovations.

They are the same move.

- A unit becomes atomic.
- Atomic becomes standardized.
- Standardization becomes infrastructure.
- Infrastructure enables emergence.

This is the pattern.

## The Semantic Container Model

I call this the Semantic Container Model.

*A semantic container is a standardized unit that carries meaning, identity, and context in a form that can be universally interpreted and reused.*

It is not defined by its medium.

It is defined by its structure:

- It contains a single, atomic unit
- It is uniquely identifiable
- It is semantically defined
- It can be linked to other containers
- It can be governed and constrained

When these conditions are met, something shifts.

The container becomes independent of the system that created it.
It can move across contexts without losing meaning.
It can connect to other containers without translation.

It becomes infrastructure.

This is why the same pattern appears across domains.

A shipping container moves cargo across ships, trucks, and rail.
A Zettelkasten note moves ideas across contexts of thought.
An RDF triple moves facts across systems.
A word, defined and shared, moves meaning across people.

Different domains.
Same structure.

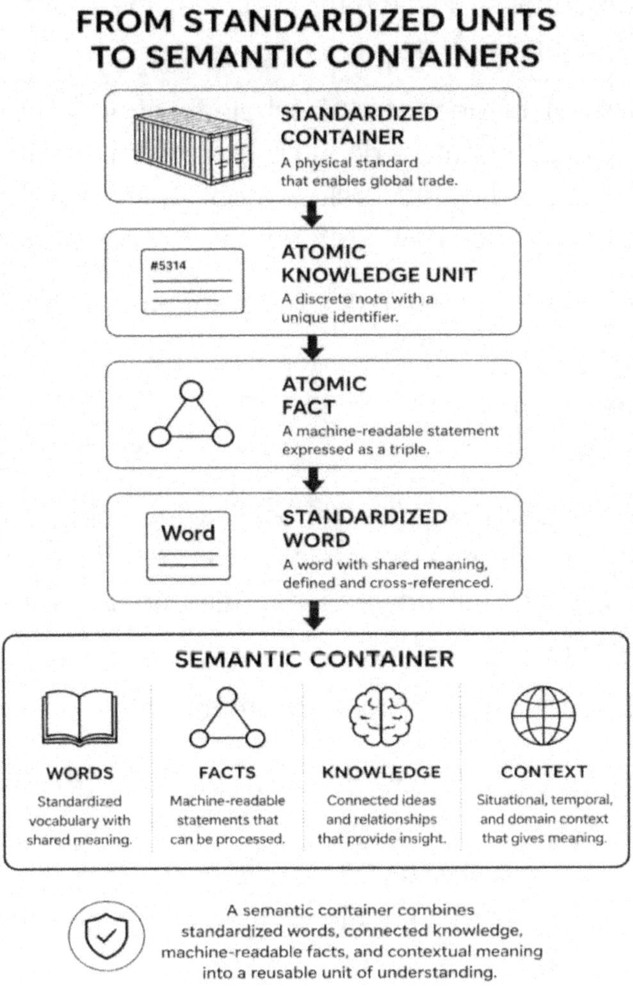


*Figure 6-27-1: From Standardized Units to Semantic Containers*

## The Missing Layer

Humanity already operates on semantic containers.

We have dictionaries.
We have standards.
We have shared language.

The Oxford English Dictionary is one of the most formidable examples of this.

It is the result of decades of scholarly effort – researchers, linguists, and contributors documenting meaning with precision and care. It is used by universities, publishers, legal institutions, and global enterprises as a reference point for language.

It works because it is trusted, consistent, and authoritative.
And yes – it is proprietary.

That is not a flaw.
It is a choice.

Oxford University Press has chosen to maintain the OED as intellectual property, accessible through subscription to institutions such as universities, law firms, and research organizations.

But that choice reveals something important.

The most complete semantic system humans rely on is:

- Not open
- Not machine-native
- Not integrated into the systems that now depend on meaning

We have built a semantic foundation for humans.

We have not built one for machines.

## The Gap

There are attempts.

Wikidata assigns identifiers and relationships at global scale. WordNet structures language into semantic relationships. Schema.org provides shared vocabularies for the web.

Each contributes.

None unify.

There is no universally accepted semantic container system for machines.

That is the gap.

And it is the reason AI struggles – not because models are insufficient, but because meaning is inconsistent.

## From Data to Meaning

AI does not operate on data alone. It operates on interpreted data.

When meaning is fragmented:

- Models infer
- Systems guess
- Outputs vary

When meaning is standardized:

- Systems align
- Reasoning stabilizes
- Outcomes become predictable

The shift is not from small models to large models.

It is from unstructured meaning to semantic containers.

# Semantic Modeling

This is where semantic modeling becomes essential.

Semantic modeling is the discipline of defining, structuring, and governing meaning so that it can be consistently interpreted across systems.

It requires:

- Clearly defined entities
- Shared ontologies
- Unique identifiers
- Explicit relationships
- Embedded governance

This is not a technical exercise. It is an architectural one.

It determines whether systems can interoperate, whether data can be trusted, and whether AI can reason. Without semantic modeling, AI remains probabilistic. With it, AI becomes operational.

By now, the pattern is clear. Not because it was explained. Because it has already been demonstrated.

Across shipping. Across knowledge management. Across the Semantic Web. Across language itself. Each example solved a different problem. Each relied on the same principle.

- Standardize the unit.
- Connect the relationships.
- Enable reuse.

Semantic Containers apply that principle to enterprise information.

They provide a standardized, governed unit capable of carrying meaning across systems, organizations, and AI agents.

But one important question remains.

If we can consistently represent reality...

**how do we consistently govern action?**

Representation tells us what exists.

It does not determine what may happen next.

That is the challenge of Transactional Governance.

## The Pattern Has Been Discovered

We are not at the beginning of this journey. We are in the early stages.

The pattern has already been discovered—multiple times, across multiple domains.

What remains is to apply it intentionally.

Where meaning is ambiguous, define it.
Where identifiers are inconsistent, standardize them.
Where relationships are implicit, make them explicit.
Where governance is external, embed it.

Because in an AI-driven world, organizations will not compete on the amount of data they possess.

They will compete on the quality of meaning they represent...

**and the quality of decisions they can safely execute.**

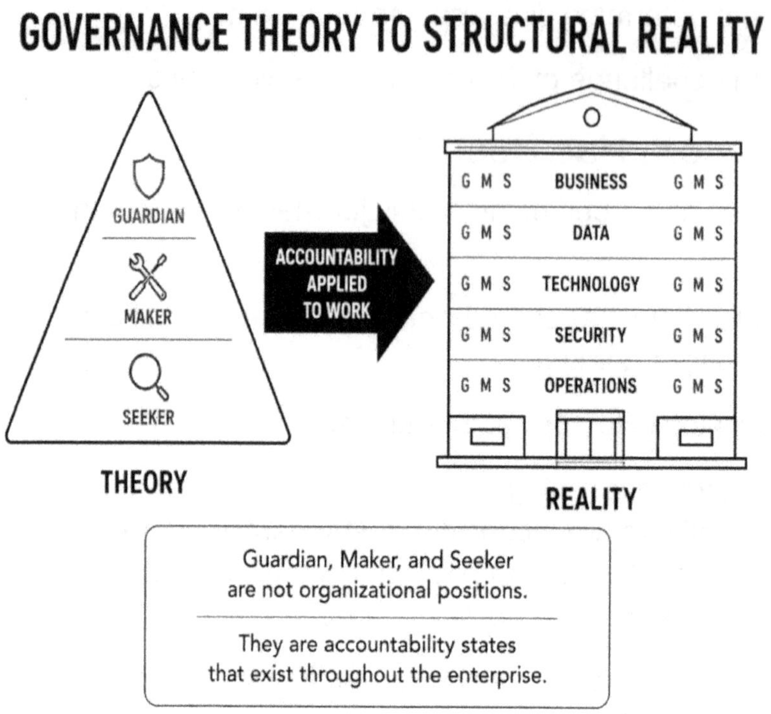


*Figure 6-28-1: Theory to Reality*

# Chapter 28: The Work That Remains

By now, the pattern is clear.

Not because it was explained – but because it has already been demonstrated.

Across domains. Across decades. Across systems.

The unit changes.
The structure holds.

And with it comes responsibility.

Throughout this book, a set of implicit contracts has been established.

- If meaning is to scale, it must be defined.
- If systems are to interoperate, identifiers must be consistent.
- If AI is to reason, relationships must be explicit.
- If trust is to be maintained, governance must be embedded – not external.

These are not philosophical positions.

They are operational requirements.

## The Shift We Cannot Ignore

For decades, governance has lived outside the moment of action.

Policies were written.
Standards were documented.
Controls were reviewed after the fact.

But decisions were made somewhere else.

That model does not survive AI.

In an agentic environment, decisions happen continuously – at machine speed, across systems, without pause. There is no moment to "check compliance later."

Governance must exist **at the point of transaction**.

Not as oversight. As structure.

Not as documentation. As execution.

This is the shift.

## The Role of GMS

This is where the Guardian-Maker-Seeker model becomes essential.

Not as a framework to understand governance – but as a model to operationalize it.

The Guardian defines and protects meaning.
The Maker encodes that meaning into systems.
The Seeker acts on that meaning in real time.

These roles do not exist in sequence.

They exist **simultaneously** – within the same transaction.

When a decision is made:

- The meaning must already be defined
- The constraints must already be embedded
- The relationships must already be known

That is governance.

Not review.
Not escalation.
Execution.

## Where Semantic Containers Fit

Semantic containers make this possible.

They carry:

- The definition of meaning
- The identity of the object
- The relationships that give it context
- The constraints that govern its use

When a system interacts with a semantic container, it is not interpreting meaning.

It is executing against it.

This is the connection.

Semantic Containers → Governed Meaning
GMS → Operational Roles
Together → Governance at the point of transaction

What Must Be Built

The work ahead is not abstract.

It is architectural.

→ Define your core entities.
→ Standardize your identifiers.
→ Establish shared ontologies.
→ Make relationships explicit.
→ Embed governance into execution pathways.
→ And align those efforts to clear accountability: Who defines meaning? Who encodes it? Who acts on it?

Build semantic containers for your data.
Operationalize them through the Guardian-Maker-Seeker framework.

## The Reality

Most organizations are not structured this way today. What we find most often is that Meaning is fragmented, Identifiers are inconsistent, Relationships are implicit, and Governance is external.

AI is already operating on top of that foundation.

The Risk is not that AI will fail. **The risk is that it will succeed on inconsistent meaning.**

## The Opportunity

This is the moment to change everything to future-proof the operation.

Not incrementally – Structurally.

The organizations that lead will not be those that deploy the most advanced models.

They will be those that can assert:

- What their data means
- How it can be used
- Who is accountable for it
- And how that accountability is enforced in real time

That is governance at the point of transaction.

A Note on Execution

This book was written to make these ideas accessible.

You can begin, now. You can define your first semantic container. You can align your first domain. You can establish clarity where ambiguity exists.

And you should. Because progress does not require perfection. But scale requires structure.

## Transactional Governance

There is a difference between understanding this model and implementing it at scale.

Designing governance that operates inside transactions requires:

- Alignment across executive leadership
- Integration into architecture and engineering workflows
- Clear ownership across Guardian – Maker – Seeker roles
- Translation of policy into executable constraints

This is where most organizations stall.

Not in awareness – in operationalization.

Remember the AI FUTURE mnemonic. These are easy steps you can begin right away.

The **AI FUTURE** mnemonic provides a structured way to think about this shift. It does not describe a sequence of projects. It describes a **set of conditions** that must exist simultaneously for AI to function responsibly at scale.

- **A – AI-First Thinking**
  AI is no longer an enhancement layered onto existing processes. It is a design assumption that reshapes how

work is conceived, decisions are made, and value is created.

- **I – Innovation Through Fluid Models**
  Static operating models cannot keep pace with systems that learn, adapt, and recombine information continuously. Innovation now depends on flexibility in roles, workflows, and authority.
- **F – Foundations Aligned to Business Strategy**
  AI readiness is not a technology roadmap. It is a business commitment. Without strategic alignment, AI initiatives fragment into experiments that never scale.
- **U – Underlying Data Maps and Marketplaces**
  AI depends on shared understanding and reusable access to information. Without mapped meaning and discoverable interfaces, intelligence remains brittle and siloed.
- **T – Trust Built Through Performance**
  Trust cannot be mandated through policy alone. In AI systems, trust is earned when outcomes are explainable, repeatable, and defensible under pressure.
- **U – Upskilling Talent and Redefining Roles**
  AI changes what people do, not whether they matter. Organizations must move beyond static job descriptions toward activity-based responsibility and continuous learning.
- **R – Resolving Fear Through Leadership**
  Resistance to AI is not irrational. It reflects legitimate concern about accountability, displacement, and loss of

control. Leadership must address these concerns directly rather than dismiss them.

- **E – Empowering Technology Teams to Architect Success**
  Architecture does not emerge from command-and-control. It emerges when technologists are trusted to design systems that reflect business intent, risk tolerance, and ethical boundaries.

## The Path Forward

Civilization has always advanced by standardizing its most important units.

We did it for goods – the shipping container.
We did it for ideas – the atomic knowledge unit.
We did it for facts – the atomic fact model of the RDF Triple.
We did it for language – the standardized word model.

Now we must do it for data – the GMS governance framework.

And we must govern it where it matters most: At the moment data is used.

Because in an AI-driven world, the question is no longer:

Do you have data? Do you have models?

It is this:

Can you assert meaning,

enforce constraints,

and trace accountability –

at the point of transaction?

# Part VII – Agentic Executable Governance

Governance has spent decades describing what organizations should do. Policies define expectations. Frameworks establish responsibilities. Committees provide oversight. Standards document controls. Each has an important role. But as decisions move from human workflows into intelligent systems, description is no longer enough. Governance must be capable of operating at the same speed, scale, and point of action as the systems it governs.

The Guardian–Maker–Seeker Framework established a different foundation for accountability: responsibility follows action rather than title. Guardians establish the boundaries within which action may occur. Makers create, change, and operate. Seekers access, analyze, and use. Together, these accountability states make it possible to determine who – or what – is responsible as work moves across people, data, applications, models, and autonomous agents.

But defining responsibility does not ensure that responsibility will be honored. A policy that cannot influence an action when it occurs remains dependent on interpretation, memory, review, and intervention. In an environment increasingly populated by autonomous and semi-autonomous systems, that gap becomes untenable. Governance must move from something we periodically verify to something the enterprise can continuously execute.

Agentic Executable Governance (AEG) is the operationalization of that shift. It moves governance from documented intent toward governed execution: translating policy, authority, context, risk, and accountability into mechanisms capable of evaluating actions at the point where they occur. The objective is not to eliminate human judgment or replace existing governance disciplines. It is to make their intent operational wherever decisions are being made.

This is where the ideas developed throughout The Cracked Egg converge. Accountability becomes executable. Policy becomes actionable. Architecture becomes a mechanism for control. Transactions produce evidence. And trust is no longer based primarily on the existence of policies, committees, or attestations, but on an organization's ability to demonstrate that its systems consistently operate within the boundaries it has established.

Part VII moves from governance framework to operational standard. It examines what must exist for governance to function at the point of action, how those capabilities work together, and ultimately how an organization can demonstrate – not merely claim – that its AI operates within governed boundaries.

That is the promise of Agentic Executable Governance:

Governance does not simply surround the action. Governance becomes part of the action.

# Chapter 29: From Governance Framework to Operational Standard

Governance is no longer something we review. It is something we build into the moment of transaction, where decisions are made and data is used. Transactional governance will take us into the AI Future.

For decades, governance has been treated as an organizational function.

Policies were written. Committees reviewed them. Controls were documented. Audits verified that procedures had been followed.

This approach worked reasonably well when technology operated at human speed. Decisions could wait for approval. Exceptions could be escalated through management chains. Governance existed outside the systems performing the work.

Artificial intelligence changes that assumption.

AI systems no longer wait for governance to catch up. They make recommendations, initiate workflows, retrieve information, coordinate actions, and increasingly collaborate with other intelligent systems. Every transaction occurs in milliseconds. Every decision has the potential to influence thousands of downstream outcomes.

The legacy model cannot respond quickly enough because it was never designed to operate inside the transaction itself.

Throughout this book we have reached the same conclusion repeatedly from different perspectives: **Governance must move to the point of execution**.

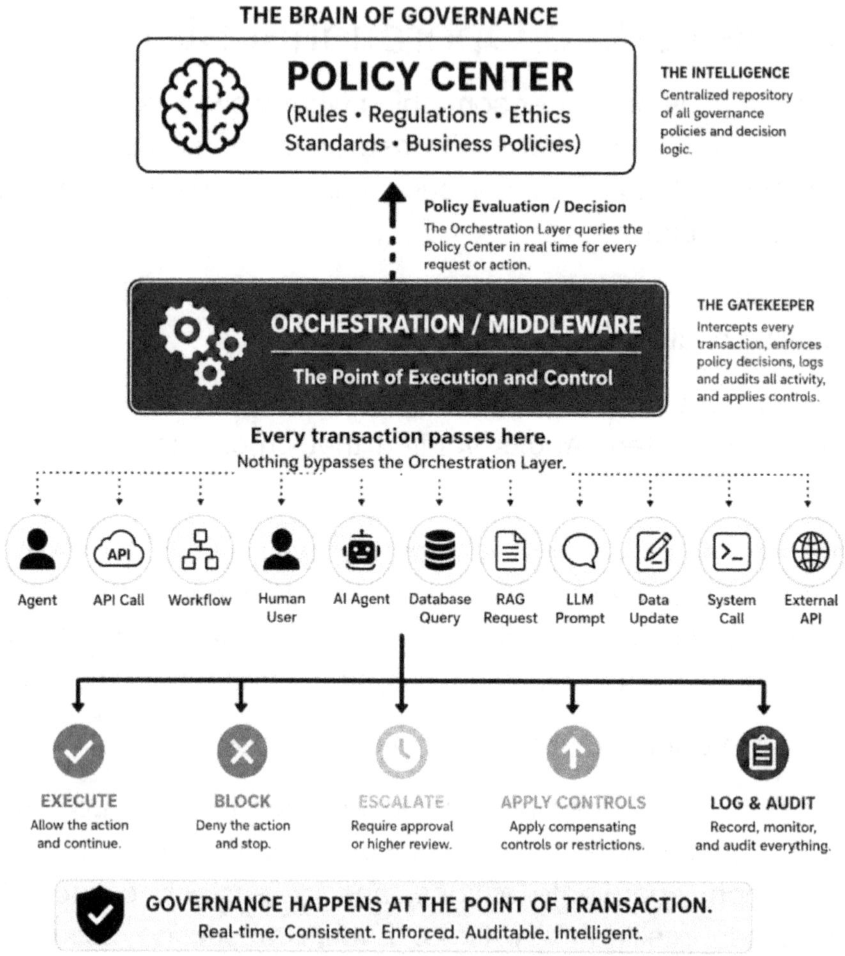


*Figure 7-29-1: The Orchestration View of the Governance Framework*

That conclusion requires more than improved policy. It requires a fundamentally different operating model.

That operating model is Agentic Executable Governance.

## Governance as Infrastructure

Agentic Executable Governance (AEG) is not another governance framework.

It is not another committee structure. It is not another maturity model.

It is an operational architecture that embeds governance directly into the systems performing work.

Rather than asking people to remember policies, AEG enables systems to evaluate policy automatically before actions occur.

Rather than documenting accountability after the fact, AEG records accountability as transactions occur.

Rather than separating governance from execution, AEG unifies them.

Governance becomes infrastructure.

Invisible to users.

Constantly active.

Continuously auditable.

This is the shift from governance as documentation to governance as execution

Agentic Executable Governance (AEG) is an operational standard.

It describes an operating model for controlling actions involving data. AEG governs what actions may be taken, by whom, under what conditions, and within what limits.

Governance becomes infrastructure.

The first component of that infrastructure is the Policy Center.

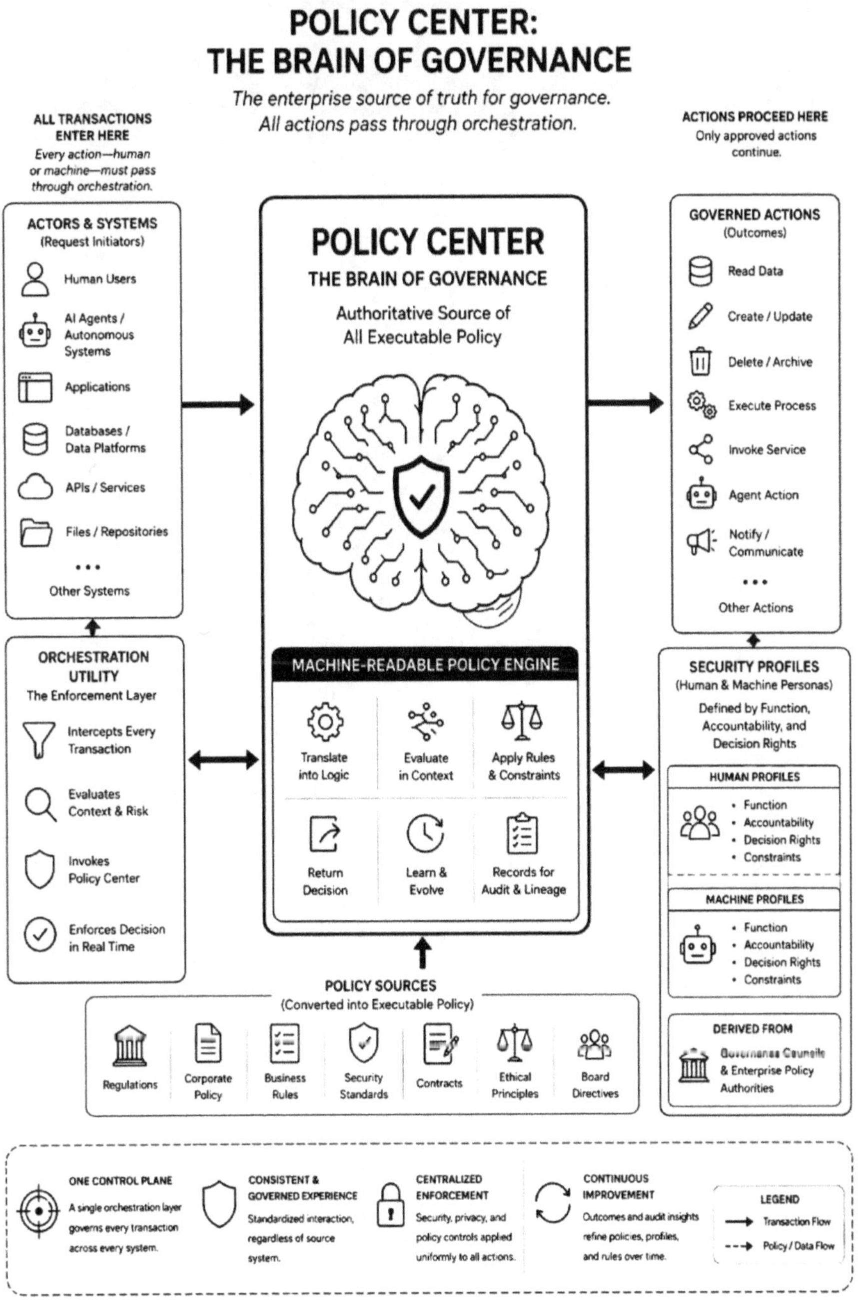


*Figure 7-30-1: The Policy Center*

# Chapter 30: The Policy Center

Every organization already practices some form of governance.

It exists in regulations, corporate policies, security standards, contracts, privacy requirements, ethics statements, and business rules.

Unfortunately, these assets exist primarily as documents written for human interpretation. The Policy Center transforms these human-readable artifacts into machine-interpretable rules. Policies become executable.

Instead of asking, *"Does this transaction comply?"*

The system asks, *"Which policies apply, and do they permit this action?"*

Policy is no longer passive. It becomes operational.

The Policy Center may be a new artifact within your organization. Companies that have mature knowledge management practices are more likely to have collected, collated, and cross-referenced their policies. This may also have been done by a mature data governance team when building the catalog and ontology. The truth is likely far from ideal for most organizations. Policies may be in a repository. But they are rarely machine readable and often locked down under a security framework the constrains employees from access. This must change in the AI era.

The transition to machine-readable representations of business processes, regulatory requirements, ethical choices, and strategic direction will require that effort is allocated to developing a strong policy center. This may be done in several ways and in partnership with activities such as ontology building or knowledge graph representation. But ultimately, they must be architected intentionally to be available to both human and Agentic actors.

These policies are not passive documents. They are business and system processes and rules translated into executable controls that can be enforced at the point of action.

The Policy Center defines constraints on:

- data use and sharing
- privacy and access rights
- interoperability requirements
- role-based security protocols
- permitted and prohibited actions
- escalation and exception handling

In this model, governance is no longer something reviewed after the fact. It becomes an operational control layer that enforces policy directly within automated and agentic systems.

Humans and AI systems are governed by the same executable rules. AI Governance is not separate from Data Governance. They must function within the same environment. What used to be manual processes and business procedures are now subsumed into the Orchestration components.

The orchestration layer determines

- what actions
- under what conditions
- with what level of impact

Governance is not binary. Executable governance evaluates graduated authority rather than simple permission. Different users and intelligent agents may perform the same activity under different authority levels, risk tolerances, approval requirements, and operational constraints.

## Policy Center and Orchestration Model within AEG

The Policy Center is the core control structure of Agentic Executable Governance (AEG). It serves as the primary interaction point for human-in-the-loop governed transctions and the authoritative source of all executable policy.

Within the Policy Center, security profiles are defined for both human and machine personas. These profiles are not role titles – they are derived from function, accountability, and decision rights as established by governance councils and enterprise policy authorities.

The Policy Center contains the machine-readable policies used by orchestration.

These policies are derived from:

- regulations
- corporate policy
- business rules
- security standards
- contracts
- ethical principles
- board directives

Rather than existing only as documents, policies become executable decision logic that the middleware evaluates in real time.

The Policy Center lies at the center of a lightweight Orchestration Utility, which acts as the enforcement layer for all actions.

All transactions – whether initiated by a human or an agentic system – must pass through this orchestration layer.

## Policy Enforcement at the Point of Action

AEG ensures that all forms of data interaction are governed through the same mechanism, including data movement (ETL / ELT), data transformation, system updates and edits, and data access and retrieval

Every action is evaluated against the specifications defined in the Policy Center.

If an action is disallowed the system returns an error, or provides permitted alternative actions, where applicable

This ensures that both human and agentic personas operate within the same defined boundaries.

## Human and Agentic Alignment

A core principle of AEG is that humans and AI systems are governed by the same executable rules.

This alignment creates operational stability, reduces risk exposure, increases trust in automated systems, and enables scalable automation and agentic expansion. By embedding governance into execution, organizations can safely expand automation while maintaining control. This also shifts human effort away from operational enforcement and toward higher-value activities such as policy definition, regulatory interpretation, and outcome-based innovation.

## Orchestration as the Control Plane

The Orchestration Utility functions as a unified control plane that intercepts all actions involving data, evaluates those actions against machine-readable policy, and enforces constraints in real time.

This utility provides a consistent point where actions can be evaluated before execution.

Identity is verified → Context is assembled → Roles are evaluated → Applicable policies are retrieved → Risk is assessed → Required approvals are determined.

Only then does the requested action proceed. The orchestration layer becomes the organization's governance control plane. It allows governance to scale at machine speed without sacrificing consistency or transparency.

## Knowledge Graph

AI cannot govern information it does not understand. A common semantic model is therefore essential.

Knowledge graphs, supported by ontologies and standardized metadata, provide that shared understanding.

They establish relationships across systems. They preserve business meaning. They provide traceability between data, policies, people, and AI agents.

Without semantic consistency, governance becomes fragmented. With semantic consistency, governance becomes repeatable. Meaning becomes portable across the enterprise.

## Governance Engine

The Governance Engine evaluates every transaction.

It determines whether an action should be:

- Allowed
- Denied
- Escalated
- Logged
- Approved
- Deferred
- Monitored

The evaluation considers both policy and context.

Who is acting?

What role are they performing?

Which information is involved?

What regulations apply?

What level of risk exists?

What downstream consequences could result?

Every decision becomes explainable.

Every outcome becomes auditable.

Every action becomes defensible.

## Guardian, Maker, and Seeker in Motion

Throughout this book we replaced static organizational titles with dynamic governance responsibilities.

Guardian.

Maker.

Seeker.

These are not positions.

They are functions performed during a transaction.

The same individual—or the same AI agent—may perform different roles throughout a workflow.

That distinction is essential.

Governance no longer follows organizational charts.

It follows responsibility.

By governing activities rather than titles, organizations gain flexibility without sacrificing accountability.

This enables humans and AI systems to collaborate using the same governance language.

## The enterprise source of truth for governance

The Policy Center contains the machine-readable policies used by the middleware.

These policies are derived from:

- regulations
- corporate policy
- business rules
- security standards

- contracts
- ethical principles
- board directives

Rather than existing only as documents, policies become executable decision logic that the middleware evaluates in real time.

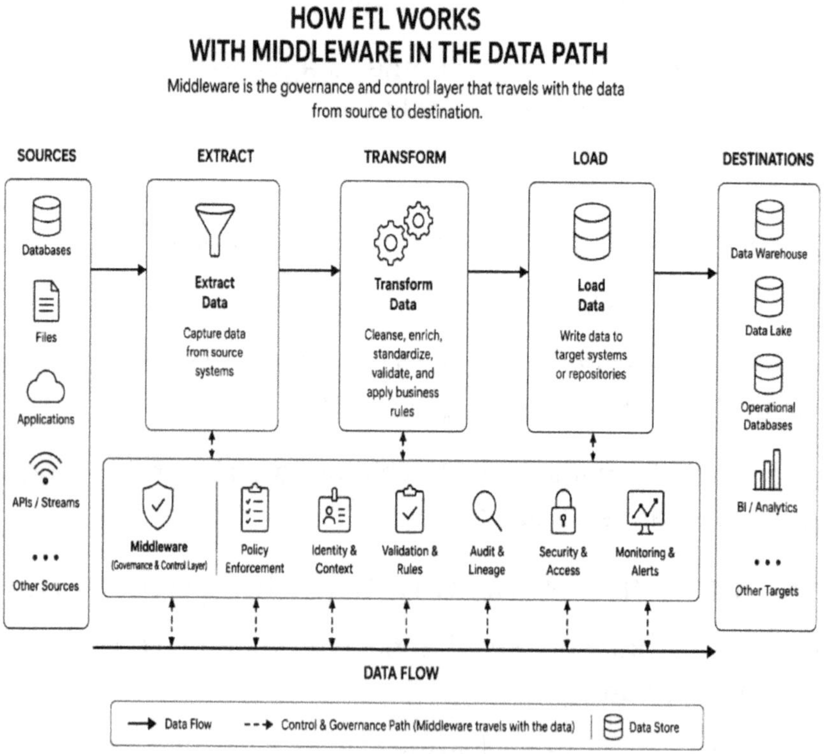


*Figure 7-31-1: How ETL Works with Middleware*

# Chapter 31: Orchestration as the Control Plane

At the center of the model is the orchestration layer held in the middleware where data actually moves and is transformed. Middleware provides the AEG Control Plane Architecture.

## Where actions originate

Every governed transaction begins with an actor requesting access to data or attempting to perform an action.

These actors may include:

- Business Applications
- AI Agents
- Databases
- APIs
- File Repositories
- Human Users

The governance challenge is not where the request comes from. It is ensuring that every request is evaluated consistently before action is taken.

The rules and guardrails that make up this first section of governance are almost always defined in project documentation that is archived, not made available for human or machine actors to reference. Changes to these rules may not be documented consistently over time.

## Middleware

Middleware is the place where governance lives

Governance should not be embedded separately inside every application.

Instead, applications submit requests through a common middleware layer that intercepts every transaction.

The middleware becomes the enterprise control plane.

Its responsibilities include:

- receiving every request
- determining transaction context
- translating data between systems
- invoking governance rules
- enforcing policy decisions
- recording audit information

Because every transaction passes through middleware, governance becomes consistent regardless of which application initiated the request.

The Orchestration Utility functions as a unified control plane that intercepts all actions involving data, evaluates those actions against machine-readable policy, and enforces constraints in real time.

This orchestration layer introduces a governed interaction layer across existing interfaces with a standardized interaction model, creating a consistent and governed user experience.

It is important to stop here for perspective. This statement, "creating a consistent and governed user experience," is highly disruptive.

Consolidation and simplification. This is the same transformation we have described throughout this book –

governance moving from external oversight into embedded execution.

Now, think of the dozens or hundreds of user interfaces in use at your company. Imagine if your software vendors could stop developing new interfaces and instead deliver their capabilities as structured data and logic, expressed through Ontologies and Knowledge Graphs, accessible by API from your company's standard user access portal, standardizing work into a single experience.

This standardization has two structural consequences:

1. It decouples the user experience from backend systems, enabling architecture and engineering teams to modernize or replace underlying technologies without disrupting users.
2. It centralizes enforcement, ensuring that security, privacy, and policy controls are consistently applied regardless of system or tool.

What appears to be a simplification of user experience is, in reality, a consolidation of control – where governance is no longer distributed across systems, but embedded in the single layer through which all actions pass.

## Authority and Action Control

The orchestration layer does not simply allow or deny actions. It enforces graduated authority levels based on persona profiles.

The level of access and capability granted to a user or agent determines:

- what actions may be taken

- under what conditions
- with what level of impact

The user experience is consistent for an executive interacting with a dashboard, or a data engineer deploying code into production, or an AI agent executing automated workflows. All are governed through the same control structure.

## Activity Defines Role – The GMS Model

Within AEG, roles are not static titles. They are defined dynamically based on activity.

**Activity → Defines → Role**

Three functional roles emerge:

**Guardian** – defines policy, enforces constraints, evaluates compliance, and provides oversight

**Maker** – performs actions that create, modify, or operationalize data, systems, or logic

**Seeker** – engages in querying, consuming, or interpreting data for decision-making

These roles are activity-based, not hierarchy-based. A single individual or system may operate in different roles depending on the action being performed.

This model ensures that governance is applied precisely at the moment of execution, aligned to the nature of the action itself.

## One Governance Model

One of the defining characteristics of Agentic Executable Governance is that humans and intelligent systems operate under the same executable governance model. Governance no longer distinguishes between whether an action originates from a person or an AI agent. It evaluates the action itself, the authority under which it is taken, the context in which it occurs, and the policies that govern it.

This consistency reduces risk, improves trust, simplifies operations, and allows organizations to expand automation without creating parallel governance models.

The time is now. You now have a blueprint to get AI Governance right. The shift to Agentic Executable Governance is not theoretical. It is architectural. It requires new patterns of design, new coordination between policy and engineering, and a redefinition of how decision authority is expressed in systems.

For many organizations, this will require guidance from those who have already begun to translate governance from documentation into executable design. Because in an agentic environment, the difference between intent and implementation is where risk and advantage are ultimately decided.

The egg has cracked, and the shell is falling away quickly. We are accountable to look past the mesmerizing glow coming from within it.

AI will soon be ever-present and integrated into everything we do. Beyond our corporate responsibilities, we have a duty to humanity to get this right.

At some point, human-in-the-loop will no longer be necessary. What remains will be the systems, constraints, and decisions we choose to design today.

The AI FUTURE will not be defined by what these systems can do.
It will be defined by what we allow them to do.

*"A framework describes what an organization should do. An operational standard ensures that it actually happens."*

# Chapter 32: The Operational Standard

For decades, organizations have approached governance as a design exercise.

Committees were established. Policies were written. Standards were published. Frameworks were approved. Audits were scheduled. Employees were trained.

Every one of these activities was intended to improve organizational behavior. Yet none of them actually governed a transaction. That distinction has become impossible to ignore.

Artificial intelligence did not create this problem. It simply exposed it.

Agentic systems do not attend governance meetings. They do not read policy manuals. They do not wait for quarterly steering committees to approve decisions. They act at machine speed, invoking APIs, transforming data, coordinating other agents, and initiating business processes long before a human can intervene.

- The governance model that served organizations for decades was built around documentation.
- The governance model required for autonomous systems must be built around execution.

That is the difference between governance as an administrative function and governance as an **operational capability.**

This distinction is the foundation of Agentic Executable Governance.

AEG is not another governance framework.

It is not another maturity model.

It is not another checklist.

It is an operational standard that evaluates every governed transaction before that action is allowed to occur.

This chapter brings together every concept introduced throughout this book into a single operational architecture.

## The Governed Transaction

Every chapter in this book has contributed one piece of a larger puzzle.

The Guardian-Maker-Seeker model established that governance is determined by activity rather than organizational hierarchy.

Semantic containers demonstrated how organizational knowledge can be represented in ways that both people and machines understand.

Ontologies and knowledge graphs provided a connected representation of enterprise reality.

Policies became machine-readable rather than human-readable.

Context transformed static rules into situational decisions.

The remaining question is remarkably simple. Everything introduced in the previous chapters now converges into a single runtime decision.

**How does all of this come together at the precise moment a transaction occurs?**

The answer is the Governed Transaction Formula.

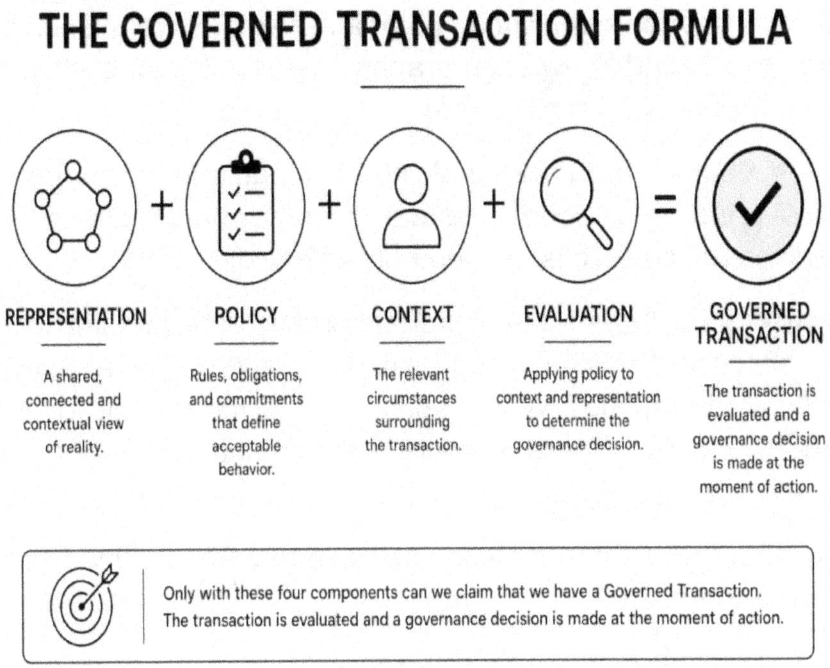


*Figure 7-32-1. The Governed Transaction Formula*

The formula appears deceptively simple.

Representation.

Policy.

Context.

Evaluation.

Together they produce something that has never existed in legacy governance – a governed transaction.

Each component performs a distinct function.

**Representation** provides a connected understanding of the business environment. Data, metadata, semantic relationships, digital twins, and knowledge graphs collectively establish what the enterprise knows to be true.

**Policy** defines organizational intent. Regulations, contracts, ethical principles, security standards, corporate directives, and business rules establish acceptable behavior.

**Context** captures the circumstances surrounding the transaction itself. Who is requesting the action? What activity are they attempting? Which asset is involved? What level of authority exists? Where is the request originating? What risks are present?

Only after these three dimensions have been assembled can governance occur.

The fourth component—**Evaluation**—is where governance actually happens.

This is the moment the organization asks a single question.

*"Given everything we know about this request, should this action be allowed?"*

Evaluation is not policy. Evaluation is the application of policy to representation within context. That distinction changes everything.

Most organizations already possess policies. Many organizations maintain high-quality metadata. Increasing numbers are investing in semantic technologies.

Very few evaluate all of those components together before action occurs.

Only when representation, policy, context, and evaluation converge at the moment of execution can we honestly claim governance has occurred.

- Everything before that point is preparation.
- Everything after that point is evidence.

Governance exists only at the point of decision.

## From Documentation to Execution

Historically, governance has been measured by artifacts.

Organizations proudly demonstrate governance through policy libraries, committee minutes, stewardship assignments, approval workflows, maturity assessments, and audit reports.

These activities remain valuable. But none of them answers the question that matters most.

**Was the transaction governed?**

That question cannot be answered by reviewing documentation. It can only be answered by examining execution.

*This represents the fundamental transition from Governance 1.0 to Agentic Executable Governance.*

The objective is no longer to demonstrate that governance policies exist. The objective is to prove that governance decisions occurred.

Every transaction.

Every time.

Regardless of whether the actor is a human employee, an application, or an autonomous AI agent. That shift transforms governance from an administrative discipline into an operational capability.

# Chapter 33: The Enterprise Control Plane

The Governed Transaction Formula explains *what* must occur.

The next question is equally important.

**Where does it occur?**

The answer is the AEG Orchestration Utility. – The *Enterprise Governance Control Plane*

## The AEG Orchestration Utility

The AEG Orchestration Utility represents the operational heart of the standard.

Rather than embedding governance independently within hundreds of disconnected applications, the organization establishes a single governance control plane through which governed transactions are evaluated.

Every request enters through the same architectural gateway.

Whether the request originates from a customer portal, a business application, an autonomous AI agent, a workflow engine, or an API, the process begins in exactly the same way.

- The request is intercepted before execution.
- Identity is resolved.
- The activity is classified.
- Relevant policies are retrieved.
- Context is assembled.
- Authority is evaluated.
- A governance decision is made.

Only then is execution permitted.

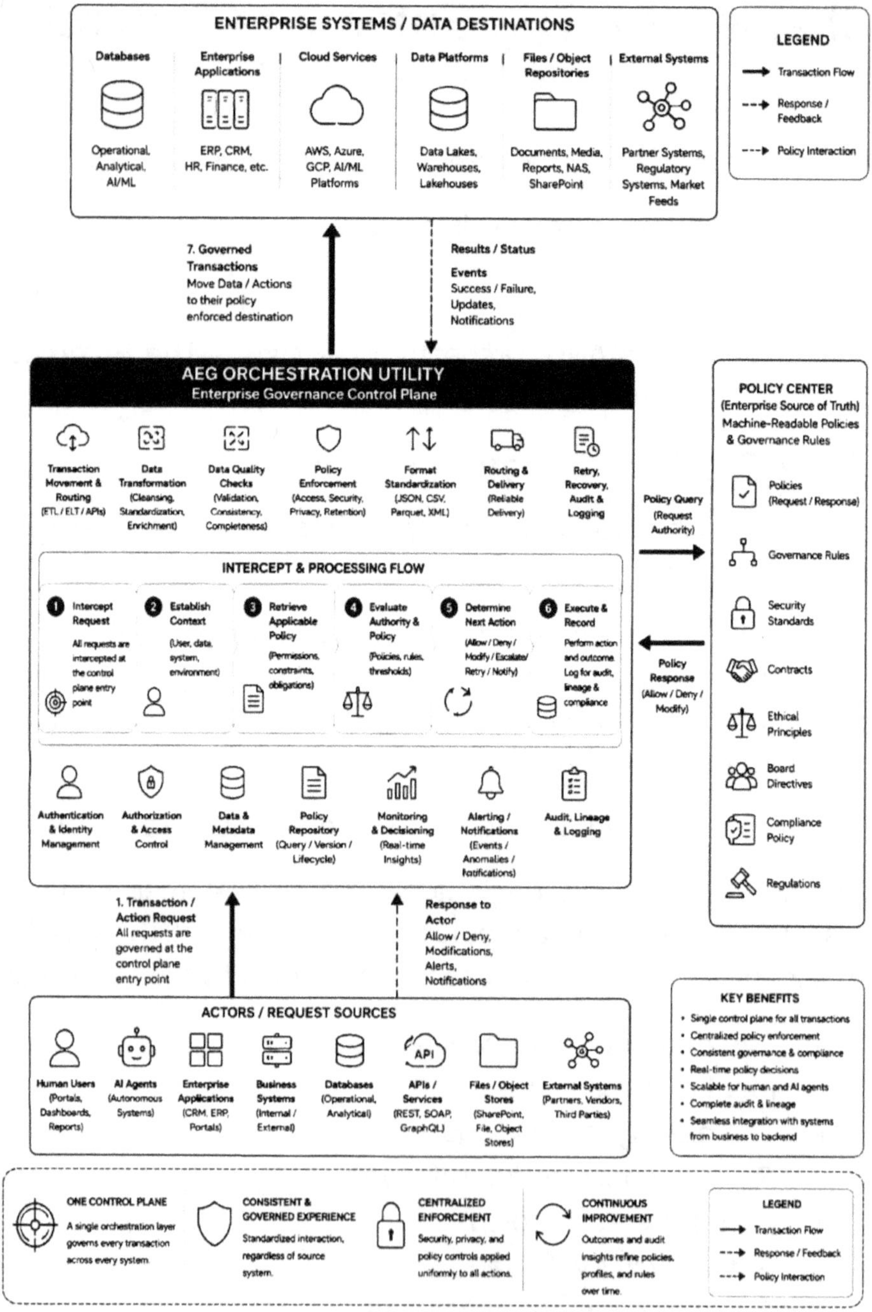
ENTERPRISE SYSTEMS / DATA DESTINATIONS
Databases
Operational, Analytical, AI/ML
Enterprise Applications
ERP, CRM, HR, Finance, etc.
Cloud Services
AWS, Azure, GCP, AI/ML Platforms
Data Platforms
Data Lakes, Warehouses, Lakehouses
Files / Object Repositories
Documents, Media, Reports, NAS, SharePoint
External Systems
Partner Systems, Regulatory Systems, Market Feeds
LEGEND
Transaction Flow
Response / Feedback
Policy Interaction
7. Governed Transactions
Move Data / Actions to their policy enforced destination
Results / Status
Events
Success / Failure, Updates, Notifications
AEG ORCHESTRATION UTILITY
Enterprise Governance Control Plane
Transaction Movement & Routing (ETL / ELT / APIs)
Data Transformation (Cleansing, Standardization, Enrichment)
Data Quality Checks (Validation, Consistency, Completeness)
Policy Enforcement (Access, Security, Privacy, Retention)
Format Standardization (JSON, CSV, Parquet, XML)
Routing & Delivery (Reliable Delivery)
Retry, Recovery, Audit & Logging
INTERCEPT & PROCESSING FLOW
1 Intercept Request
All requests are intercepted at the control plane entry point
2 Establish Context
(User, data, system, environment)
3 Retrieve Applicable Policy
(Permissions, constraints, obligations)
4 Evaluate Authority & Policy
(Policies, rules, thresholds)
5 Determine Next Action
(Allow / Deny / Modify / Escalate/ Retry / Notify)
6 Execute & Record
Perform action and outcome. Log for audit, lineage & compliance
Authentication & Identity Management
Authorization & Access Control
Data & Metadata Management
Policy Repository (Query / Version / Lifecycle)
Monitoring & Decisioning (Real-time Insights)
Alerting / Notifications (Events / Anomalies / Notifications)
Audit, Lineage & Logging
Policy Query (Request Authority)
Policy Response (Allow / Deny / Modify)
POLICY CENTER
(Enterprise Source of Truth)
Machine-Readable Policies & Governance Rules
Policies (Request / Response)
Governance Rules
Security Standards
Contracts
Ethical Principles
Board Directives
Compliance Policy
Regulations
1. Transaction / Action Request
All requests are governed at the control plane entry point
Response to Actor
Allow / Deny, Modifications, Alerts, Notifications
ACTORS / REQUEST SOURCES
Human Users (Portals, Dashboards, Reports)
AI Agents (Autonomous Systems)
Enterprise Applications (CRM, ERP, Portals)
Business Systems (Internal / External)
Databases (Operational, Analytical)
API
APIs / Services (REST, SOAP, GraphQL)
Files / Object Stores (SharePoint, File, Object Stores)
External Systems (Partners, Vendors, Third Parties)
KEY BENEFITS
Single control plane for all transactions
Centralized policy enforcement
Consistent governance & compliance
Real-time policy decisions
Scalable for human and AI agents
Complete audit & lineage
Seamless integration with systems from business to backend
ONE CONTROL PLANE
A single orchestration layer governs every transaction across every system.
CONSISTENT & GOVERNED EXPERIENCE
Standardized interaction, regardless of source system.
CENTRALIZED ENFORCEMENT
Security, privacy, and policy controls applied uniformly to all actions.
CONTINUOUS IMPROVEMENT
Outcomes and audit insights refine policies, profiles, and rules over time.
LEGEND
Transaction Flow
Response / Feedback
Policy Interaction

*Figure 7-33-1: The Enterprise Governance Control Plane*

This architecture fundamentally changes the role of governance within the enterprise.

Instead of attempting to govern individual systems, governance governs the transaction itself.

- Enterprise applications become *execution platforms*.
- The Orchestration Utility becomes the *governance platform*.

This separation creates several important advantages.

First, governance becomes consistent across the enterprise.

The same policy engine evaluates every transaction regardless of which application initiates it.

Second, governance becomes explainable.

Every decision can be reconstructed because every input into the evaluation process has been recorded.

Third, governance becomes adaptable.

New regulations no longer require modifications to dozens of applications.

They require updates to executable policies maintained within the Policy Center.

Finally, governance becomes scalable.

As organizations introduce new AI agents, applications, and automation platforms, those systems inherit governance simply by participating in the governed transaction architecture. Governance no longer depends upon application developers remembering to implement policy correctly. The architecture enforces it.

This is why the Orchestration Utility should be viewed as an enterprise control plane rather than middleware.

Middleware moves information.

The Enterprise Control Plane governs information before movement occurs.

That distinction is profound.

One manages integration.

The other manages organizational trust.

# Chapter 34: Trust as a Continuous Operational Process

Legacy governance has always viewed trust as something that is established periodically.

- An audit demonstrates compliance.
- A committee approves a policy.
- A certification confirms a point-in-time assessment.

These events are valuable, but they are snapshots. Autonomous systems do not operate in snapshots. They operate continuously.

Consequently, trust must become a continuous operational process rather than an intermittent administrative activity.

This leads to the final architectural component of Agentic Executable Governance.

## The Operational Trust Loop

Unlike legacy governance, AEG does not view compliance as the destination. Compliance becomes one outcome of a much larger operational system. Every governed transaction produces evidence. Evidence produces measurement. Measurement identifies opportunities for improvement. Improvement refines policy. Updated policy governs the next transaction.

Trust is no longer declared.

It is continuously demonstrated.

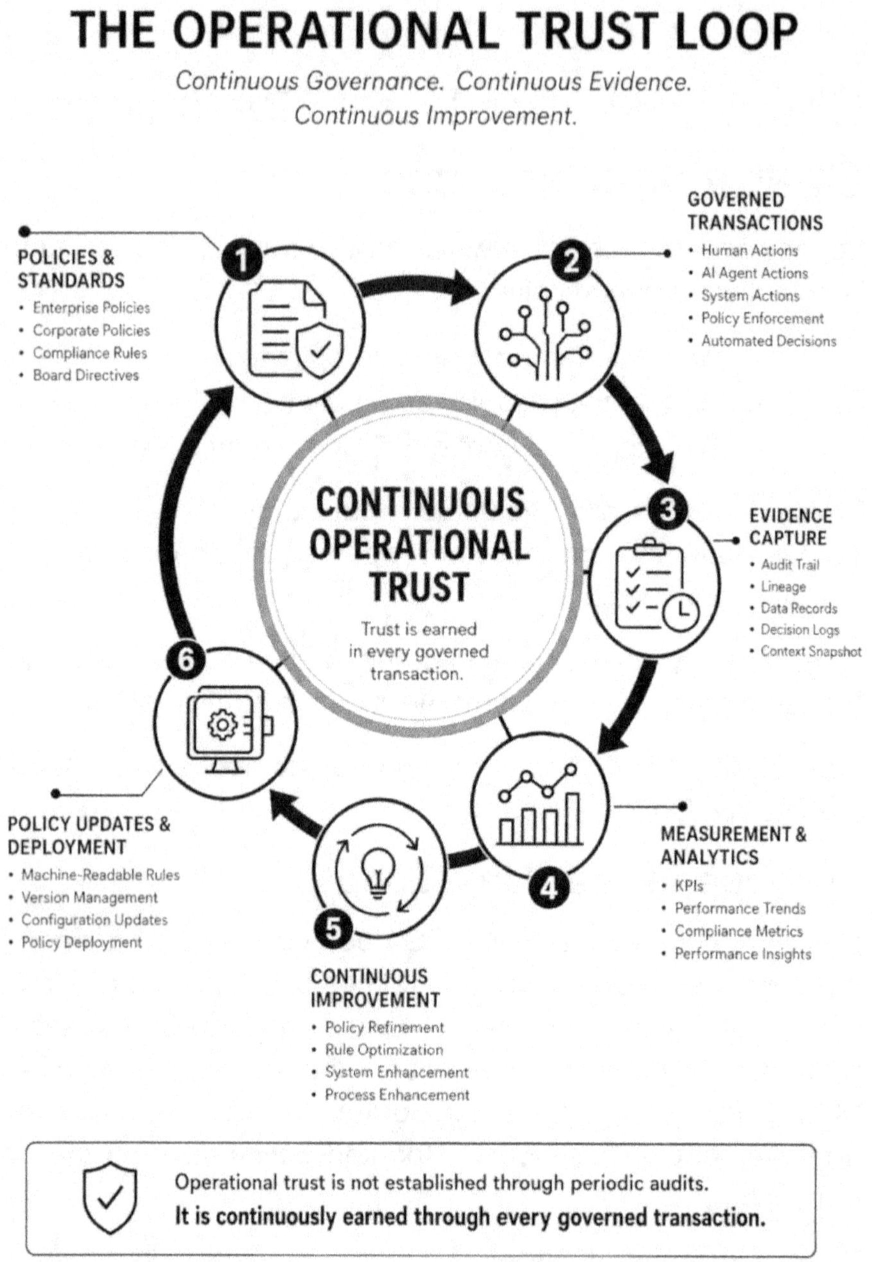


*Figure 7-34-1 The Operational Trust Loop*

## From Architecture to Standard

Every architectural component described in this chapter serves a single purpose. To ensure that governance occurs before action rather than after failure. That seemingly small shift changes the role of governance forever.

- Frameworks describe best practices.
- Architectures explain implementation.
- Operational standards establish consistent execution.

*Agentic Executable Governance* is the convergence of all three.

It defines the principles. It specifies the architecture. Most importantly, it establishes a repeatable operational model through which organizations can demonstrate that governance occurred—not as an aspiration, but as an observable, auditable, and repeatable fact. That distinction carries implications beyond technology. It changes how boards oversee risk, how regulators evaluate accountability, how auditors verify controls, and how organizations earn trust in an age of autonomous decision-making.

The next chapter explores those implications. It asks a question that extends far beyond architecture:

**If governance can now be executed, measured, and proven at the moment of action, what becomes the new standard of trust—and how will organizations demonstrate that they meet it?**

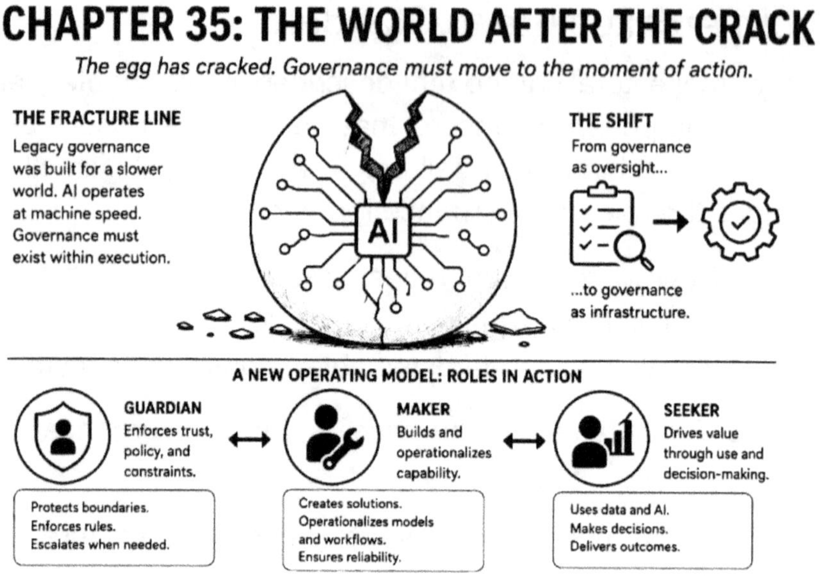


*Figure 7-35-1: The World After the Crack*

# Chapter 35: The World After the Crack

The egg has cracked.

Not shattered. Not fully opened. But fractured in a way that cannot be reversed.

What you have seen throughout this book is not a distant future. It is a shift already underway. AI is no longer operating at the edges of the enterprise. It is moving into the center – into decisions, workflows, and the precise moments where data becomes action.

And that changes everything.

For decades, organizations have built governance for a different world. A world where data moved slowly, decisions could wait, and humans reviewed outcomes before anything meaningful occurred. Governance was designed as oversight – applied after the fact through policy, process, and control structures.

That model does not survive contact with AI.

In an environment where agents act in real time and models generate outcomes instantly, governance cannot lag behind execution. It must exist within it. This is the fracture line. This is the crack in the egg.

## What Must Change

This book has advanced a single, foundational position:

**Governance must move to the moment of action.**

Not as philosophy, but as design.

To make that shift real, we introduced a new operating model grounded in function, not hierarchy:

- The **Guardian** enforces trust, policy, and constraint
- The **Maker** builds and operationalizes capability
- The **Seeker** drives value through use and decision-making

These roles are not static titles. They are **contextual responsibilities**, assumed dynamically as work is performed. Governance, therefore, is no longer assigned once. It is continuously asserted, evaluated, and enforced in real time.

This reframing resolves a critical failure in legacy governance: the assumption that accountability can be pre-defined and remain valid regardless of context. In an AI-driven system, context is everything.

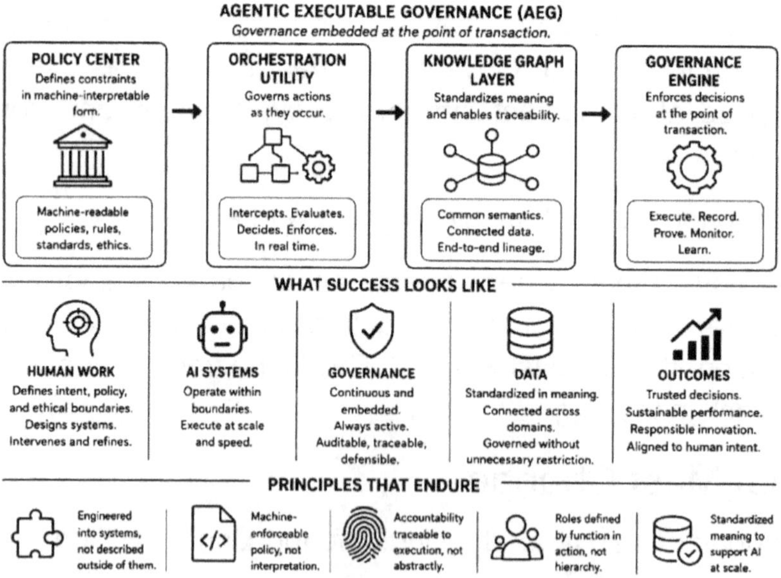


Figure 7-35-2: What comes next

To support this model, we introduced **Agentic Executable Governance (AEG)** – a system-level approach that connects policy, architecture, and execution:

- A **Policy Center** that defines constraints in machine-interpretable form
- An **Orchestration Utility** that governs actions as they occur
- A **Knowledge Graph layer** that standardizes meaning and enables traceability
- A **Governance Engine** that enforces decisions at the point of transaction

This is the shift from governance as documentation to governance as infrastructure.

Where We Are Today

Despite rapid advances in AI, most organizations remain early in this transformation.

Data is still fragmented. Definitions are inconsistent. Policies exist but are disconnected from system behavior. Governance is performed through review rather than enforcement. AI capabilities are being introduced faster than the organization's ability to control them.

There are pockets of maturity – well-managed data products, clean domains, isolated successes. But they are not yet systemic.

The reality is straightforward:

**We are in the earliest stages of cracking the egg.**

This is not a critique. It is a condition of the moment. Every major technological shift begins this way – uneven, exploratory, and often misunderstood.

What matters is not perfection. It is direction.

## The Decision Point

This moment presents a clear choice.

Organizations can extend existing governance models – adding more policies, more reviews, more layers of oversight – in an attempt to control systems that now operate at machine speed.

Or they can accept the structural shift and **rebuild governance as an embedded, executable capability**.

Only one of these paths will scale.

The difference between leaders and laggards in the AI era will not be defined by access to models. Those are becoming commoditized. It will be defined by the ability to **control how those models operate within the enterprise**.

That is a governance problem. And it is now an engineering problem.

## Principles That Endure

While there is no single blueprint for success, there are principles that will hold across organizations, industries, and technologies:

- Governance must be **engineered into systems**, not described outside of them
- Policy must be **machine-enforceable**, not dependent on human interpretation

- Accountability must be **traceable to execution**, not assigned abstractly
- Roles must be **defined by function in action**, not position in hierarchy
- Data must have **standardized meaning** to support AI at scale

The EZ-AI-Ready GMS Framework and AEG model bring these principles together into a cohesive design. They connect business outcomes, technical architecture, and governance into a single operating system for AI.

They are not the only approach.

But they represent a **complete approach** – one that addresses the business problem, the system design, and the human model simultaneously.

## What Success Looks Like

If we implement these principles effectively, the future is not one of diminished human relevance. It is one of elevation.

**Human** work shifts upward - defining intent, policy, and ethical boundaries. Designing systems that enforce those boundaries. Intervening when escalation is required. Continuously refining the rules that govern AI behavior.

**AI systems** operate within those boundaries - executing decisions at scale and speed. Applying consistent logic across scenarios. Adapting within predefined constraints.

**Governance** becomes continuous - Embedded in every transaction. Invisible to the user, but always active. Auditable, traceable, and defensible.

**Data** becomes usable - Standardized in meaning. Connected across domains. Governed without restricting access unnecessarily.

This is not a theoretical end state. It is a practical direction – one that can be built incrementally but must be designed intentionally.

## The Responsibility Ahead

With this shift comes responsibility.

We are not simply adopting new technology. We are defining how decisions are made, how data is used, and how trust is maintained in an AI-driven world.

Those decisions will not be made solely by regulators, vendors, or centralized governance bodies.

They will be made by the people who design and operate these systems.

That includes executives setting direction, architects designing platforms, engineers building capabilities, and practitioners using data to drive outcomes.

The responsibility is shared. And it is immediate.

Because once systems are in place, they shape behavior. They define what is possible, what is allowed, and what is prevented.

In that sense, governance is no longer a control function.

It is a **design function**.

## The Cracked Egg

The cracked egg is not a warning.

It is an inflection point.

An opportunity to move beyond governance as theater – policies that describe intent without enforcing it – and toward governance as orchestration, where control is embedded directly into the systems that act.

The future of AI will not be determined by the sophistication of models alone. It will be determined by the systems that govern them. And those systems are not inevitable. They are built.

Deliberately. Intentionally. Architecturally.

That is the work in front of us.

**"Governance is no longer something we review. It is something we build into the moment of execution – where decisions are made, constraints are enforced, and AI systems act within defined authority."**

That is how organizations move from experimentation to control.

That is how trust is sustained at scale.

That is how we ensure that the AI future is not only powerful – but responsible, transparent, and aligned to human intent.

The egg has cracked.

What comes next is up to us.

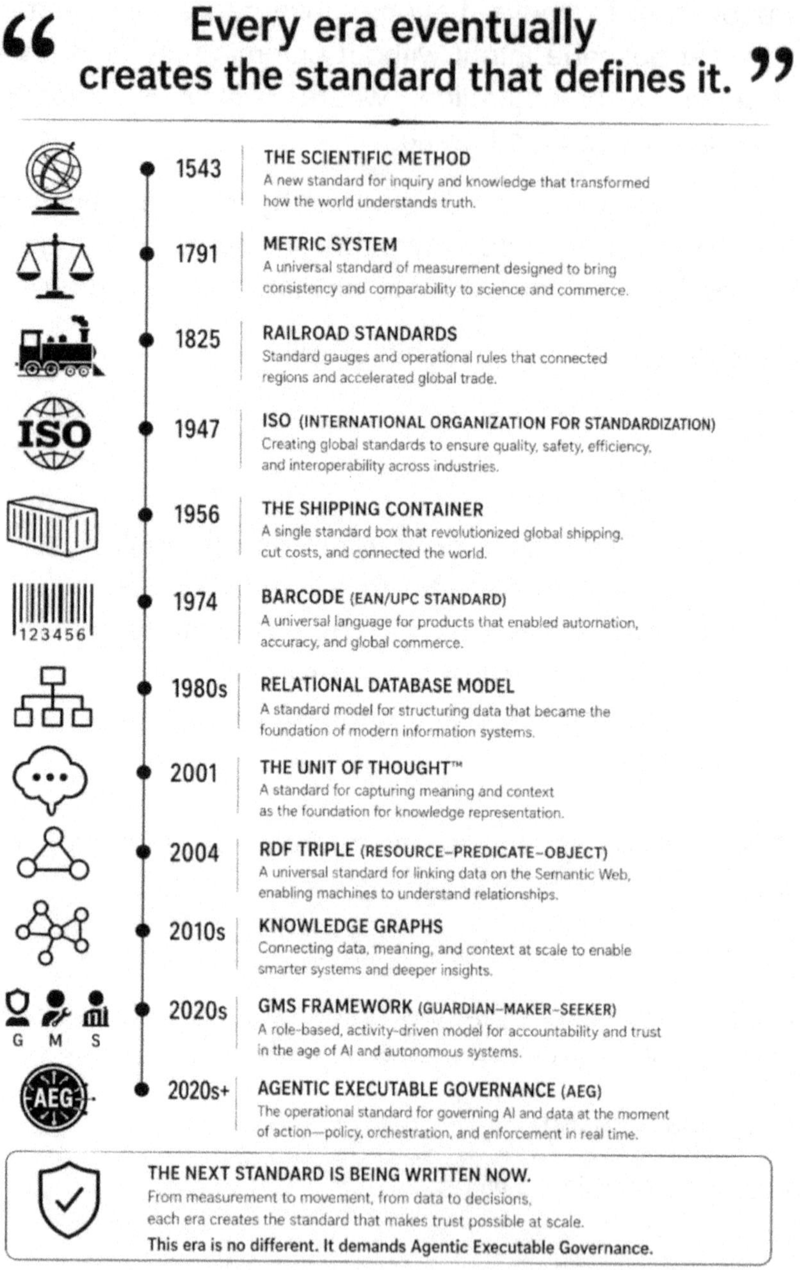


*Figure 8-0-1: Every Era Defines the Standards that Define It*

# Part VIII: The AEG Certification Standard

## From Operational Standard to Global Trust

The egg has cracked.

Not because artificial intelligence became intelligent. Not because large language models learned to write. Not because autonomous agents began completing work once reserved for people.

The egg cracked because, for the first time in history, intelligence became capable of acting without waiting for human permission.

For centuries, governance was designed around an assumption that no longer holds true. People acted. Organizations documented. Auditors verified. Executives accepted accountability. The pace was measured in weeks, months, and sometimes years.

Artificial intelligence has changed none of those responsibilities. It has simply compressed the timeline. Decisions that once unfolded over weeks now occur in milliseconds. Thousands of governed transactions become millions. Millions become billions.

Governance itself has not become obsolete. It has become the critical infrastructure of trust.

Throughout this book, we have explored why governance must evolve from paperwork into execution, from committees into operational capability, and from documentation into continuous evaluation.

Those ideas culminated in Agentic Executable Governance.

But AEG is not the destination. It is the beginning.

# Chapter 36: Every Transformation Creates a Standard

History follows a remarkably consistent pattern.

Innovation always arrives before standardization.

- The automobile came before traffic laws.
- Electricity came before electrical codes.
- Accounting existed long before Generally Accepted Accounting Principles.
- The Internet flourished before cybersecurity frameworks matured.
- Cloud computing spread years before organizations developed mature cloud governance.

Every transformational technology eventually reaches the same inflection point.

→ The technology becomes common

→ The risks become visible

→ Markets lose confidence

→ Society demands consistency

→ A standard emerges.

The organizations that embrace that standard become trusted. The organizations that ignore it eventually become cautionary tales. Artificial intelligence is now approaching that same moment.

Today, organizations compete by announcing how much AI they have deployed. Tomorrow, they will compete by proving that it can be trusted.

Those are not the same thing.

One measures capability. The other measures confidence. History consistently rewards confidence.

## Beyond Governance Frameworks

For decades, governance has largely been described through frameworks.

- Frameworks organize thinking.
- They define responsibilities.
- They identify capabilities.
- They recommend best practices.
- They provide valuable guidance.

They answer the question,

> *"What should an organization do?"*

Operational standards answer a fundamentally different question.

> *"Can an organization demonstrate that it actually did it?"*

That distinction separates aspiration from execution. It is the difference between documenting accountability and proving accountability.

*Agentic Executable Governance* is offered as an operational standard.

It does not replace governance frameworks. It operationalizes them.

ISO 42001 remains an important milestone in the evolution of AI governance.

- It establishes organizational expectations.
- It encourages systematic management.
- It creates consistency in governance programs.

Those are necessary achievements. They are not sufficient for autonomous execution.

Agentic systems require governance decisions that occur before action.

The moment an AI agent invokes another agent, accesses regulated information, transforms enterprise data, executes a financial transaction, or initiates a clinical workflow, governance cannot remain a document waiting on a shelf.

It must become **executable**.

That is the distinction between governance programs and operational governance.

One describes. The other acts.

## Trust Becomes the Competitive Advantage

Organizations have traditionally competed through cost, quality, innovation, and customer experience.

The next competitive advantage will be trust.

Not marketing trust. Operational trust. The ability to demonstrate, with evidence rather than promises, that autonomous systems consistently behave within organizational intent.

- Customers will demand it.
- Regulators will require it.

- Partners will expect it.
- Boards will become responsible for it.

Operational trust becomes measurable because every governed transaction leaves evidence.

Every evaluation creates a decision record.

Every decision contributes to operational intelligence.

Every operational outcome continuously improves future governance.

Trust becomes observable.

Repeatable.

Auditable.

Scalable.

For the first time, organizations will possess an objective way to answer a question that has historically depended upon reputation alone.

*"Can we trust this AI?"*

## The New Responsibility of Leadership

The rise of autonomous intelligence changes leadership itself. Boards can no longer govern technology through periodic reviews. Executives can no longer assume that operational controls are embedded simply because policies exist. Governance professionals can no longer define success by completed documentation. The responsibility of leadership becomes operational.

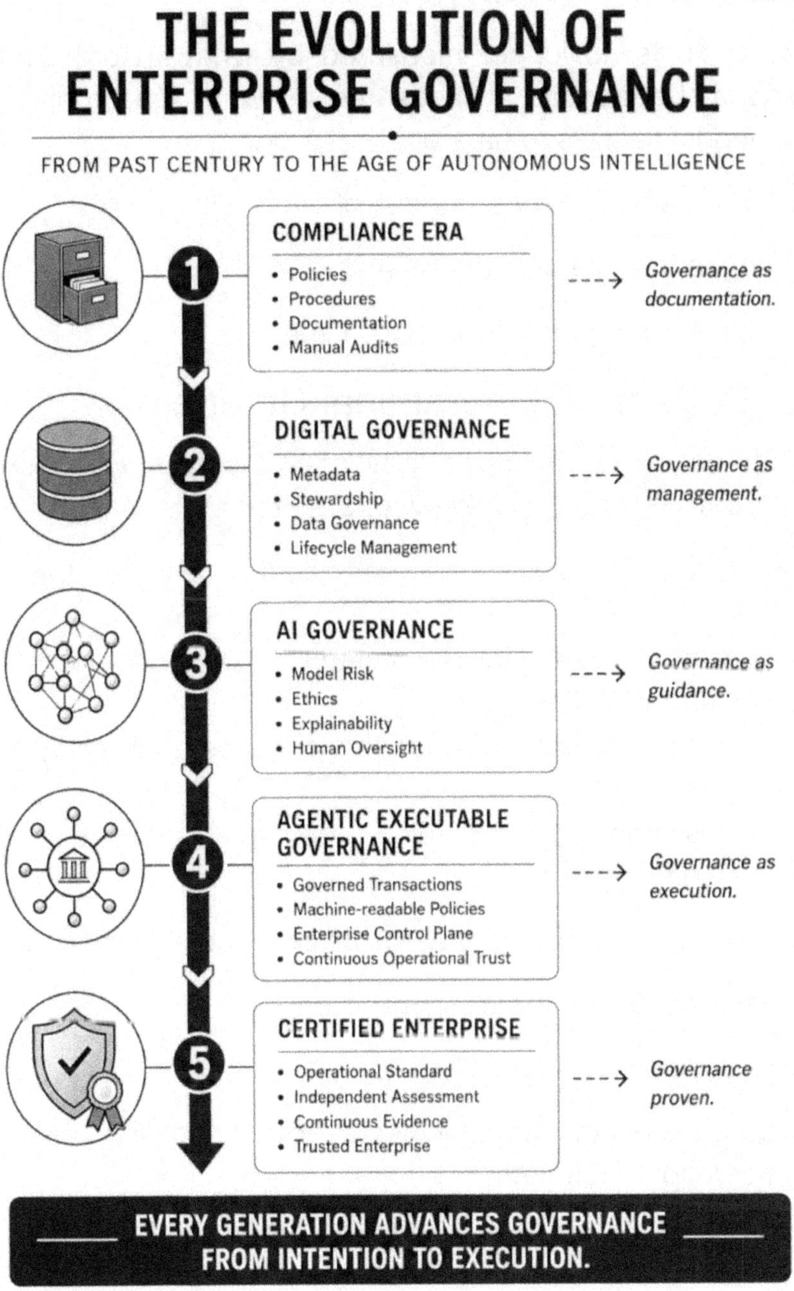


Figure 8-36-1: The Evolution of Enterprise Governance

Leadership is no longer measured by what an organization intended. Leadership is measured by what the organization continuously allows to happen.

That distinction changes the role of every executive.

It changes the role of every board.

It changes the role of governance itself.

## The Birth of an Operational Standard

Standards matter because they create confidence between organizations that do not know one another.

- A customer trusts an aircraft because aviation standards exist.
- A bank trusts financial statements because accounting standards exist.
- Manufacturers trust suppliers because quality standards exist.
- Hospitals trust medical devices because safety standards exist.

Artificial intelligence now requires its own operational standard.

Not because governments demand one.

Because markets eventually will.

The purpose of Agentic Executable Governance is to provide that operational foundation.

Not by replacing existing standards.

By connecting them.

- Regulations become executable.

- Policies become operational.
- Evidence becomes continuous.
- Governance becomes measurable.
- Trust becomes demonstrable.

## Why Certification Matters

Training teaches knowledge. Certification demonstrates competence. Operational certification demonstrates organizational capability.

These distinctions become increasingly important in the age of autonomous systems.

An executive may understand governance. A board may endorse governance. An enterprise may publish governance.

Only an independently assessed organization can demonstrate that governance consistently operates at the point of action.

That is why certification becomes the natural extension of an operational standard.

Not only as a marketing credential.

Not just another framework.

**As objective evidence.**

The marketplace has repeatedly demonstrated that trust scales when conformance becomes independently verifiable.

The future of AI governance will be no different.

Organizations will increasingly ask not whether governance exists, but whether it has been proven. AEG Certification is intended to answer that question.

THE AEG CERTIFICATION ECOSYSTEM
A HIERARCHY OF TRUST. A FOUNDATION FOR THE FUTURE.
AGENTIC EXECUTABLE
AEG
GOVERNANCE
AEG CERTIFIED BOARD™
Executive Oversight of Autonomous Enterprises
AEG CERTIFIED ENTERPRISE™
Operational Governance Proven at the Point of Transaction
GOVERNANCE
Policy Design
Oversight
Accountability
Decision Rights
ARCHITECTURE
Design Principles
Integration
Standards
Scalability
SECURITY
Access Control
Protection
Identity
Resilience
RISK & COMPLIANCE
Risk Management
Regulatory Alignment
Controls
Policy Enforcement
AUDIT & ASSURANCE
Verification
Auditability
Independence
Assurance
DATA & AI OPERATIONS
Data Governance
Quality
Lineage
AI Operations
INDIVIDUAL CERTIFICATIONS
PRACTITIONER
Foundational
Knowledge
Application
PROFESSIONAL
Applied Expertise
Implementation
Best Practices
ARCHITECT
Design Solutions
Architecture
Leadership
AUDITOR
Evaluate
Assess
Report
ADMINISTRATOR
Operate
Configure
Maintain
ASSESSOR
Assess Maturity
Validate
Improve
THE AEG OPERATIONAL STANDARD
GOVERNED TRANSACTIONS
ENTERPRISE CONTROL PLANE
MACHINE-READABLE POLICY
CONTINUOUS OPERATIONAL TRUST
EVIDENCE-BASED CONFORMANCE
CERTIFICATION DOES NOT CREATE TRUST. IT DEMONSTRATES THAT TRUST HAS BEEN OPERATIONALIZED.

*Figure 8-36-2: The AEG Certification Ecosystem*

## The Invitation

This book argues that artificial intelligence did not create the governance crisis. It revealed one that had existed all along.

It challenged assumptions that governance could remain separate from execution.

It demonstrated that context matters as much as policy, that semantics matter as much as structure, and that governance must occur at the precise moment a decision is made.

These ideas are offered not as the conclusion of a conversation but as its beginning. Operational standards are never created by a single author. They mature through practitioners. Through researchers. Through regulators. Through educators. Through architects. Through organizations willing to test them. Through leaders willing to improve them.

The work now belongs to a broader community.

## The AEG Manifesto

Every generation inherits a defining management challenge.

Frederick Taylor transformed industrial work.
W. Edwards Deming transformed quality.
Peter Drucker transformed management.
The digital age transformed information.

**Our generation must transform trust.**

Not because machines have become human.

But because human institutions must now govern machines capable of acting on our behalf.

We believe the future will not belong simply to the organizations with the largest models, the most data, or the fastest algorithms.

**It will belong to the organizations that can prove their decisions deserve to be trusted.**

Not because machines have become human.

But because human institutions must now govern machines capable of acting on our behalf.

The future will not belong to the organizations with the fastest models.

It will belong to the organizations whose decisions can be trusted.

Whose actions can be explained.

Whose policies execute automatically.

Whose accountability survives autonomy.

Whose governance is measured not by the documents they publish, but by the decisions they continuously make.

Agentic Executable Governance is offered as one path toward that future.

Not as the final answer.

But as an invitation to build something larger than any one organization, technology, or generation.

The egg has cracked.

What hatches next is no longer a question of artificial intelligence.

It is a question of human leadership.

And that future begins – not with better algorithms – but with trust.

## We Believe

**Governance must exist where action occurs.**

A policy that cannot influence a decision at the moment it is made is an intention, not a control.

**Trust must be engineered, not assumed.**

Trust does not come from the existence of a policy, committee, certification, or audit. It is earned when systems repeatedly operate within defined authority and produce evidence that they did so.

**Accountability must follow action.**

Responsibility cannot depend solely upon titles, organizational charts, or static assignments. As humans and machines move between activities, accountability must move with them.

**Policy must become executable.**

Rules written only for humans cannot govern systems operating at machine speed. Policy must be capable of being interpreted, evaluated, enforced, and evidenced through technology.

**Autonomy must never mean absence of authority.**

AI systems may act independently. They must never act without boundaries. Every autonomous action must occur within authority established by those accountable for its consequences.

**Meaning must travel with data.**

AI cannot reliably govern what the enterprise cannot consistently understand. Data must carry sufficient context, provenance, semantics, and lineage for humans and machines to know what it means and how it may be used.

**Evidence must be continuous.**

Governance cannot depend upon reconstructing what happened months after an action occurred. Every governed transaction should leave evidence of the context, policy, authority, decision, and outcome that shaped it.

**Human judgment must rise, not disappear.**

The purpose of executable governance is not to remove people from responsibility. It is to move human judgment toward the decisions where it matters most: intent, ethics, boundaries, exceptions, escalation, and accountability.

**Governance must enable value, not merely prevent harm.**

The purpose of governance is not control for its own sake. It is to create the conditions in which data, AI, people, and autonomous systems can act with greater confidence, speed, and freedom because their boundaries are known.

**Trust must be measurable.**

We must stop judging governance by the number of policies written, committees convened, controls documented, or audits completed.

The measure of governance is whether the organization behaves as intended when decisions are actually made.

## Therefore, We Commit

We will move governance from **documentation to execution**.

From periodic review to **continuous evaluation**.

From static roles to **accountability in action**.

From policies people must remember to **controls systems can enforce**.

From retrospective evidence to **evidence created with every governed transaction**.

From governance around AI to **governance operating with AI**.

From assumed trust to **demonstrable trust**.

And from asking whether an organization *has governance* to proving that governance **works**.

## The Standard Ahead

Agentic Executable Governance is offered as an operational standard for making these commitments real.

Not as the final word on governance.

Not as a claim that every architecture must look the same.

And not as a replacement for the standards, regulations, risk frameworks, security controls, and professional disciplines that brought us here.

It is a proposition that those foundations must now become operational at the point where action occurs.

Because intelligence is becoming autonomous.
Authority cannot become ambiguous.

Because decisions are accelerating.
Governance cannot remain slow.

Because machines can act.
Humans must remain accountable.

And because trust can no longer be established periodically,
**it must be continuously earned.**

## The AEG Commitment

- **Trust must be designed.**
- **Governance must execute.**
- **Accountability must survive autonomy.**
- **Policy must travel with action.**
- **Evidence must follow every governed decision.**
- **Human authority must remain visible.**
- **And technology must operate within the boundaries humanity chooses.**

The egg has cracked.

What hatches next is no longer simply a question of artificial intelligence.

It is a question of human leadership.

**We will not govern the AI future by documenting what we hoped would happen.**

**We will govern it by building systems that make trustworthy action possible.**

That is the standard before us.

That is the responsibility we inherit.

**This is the future we choose to build.**

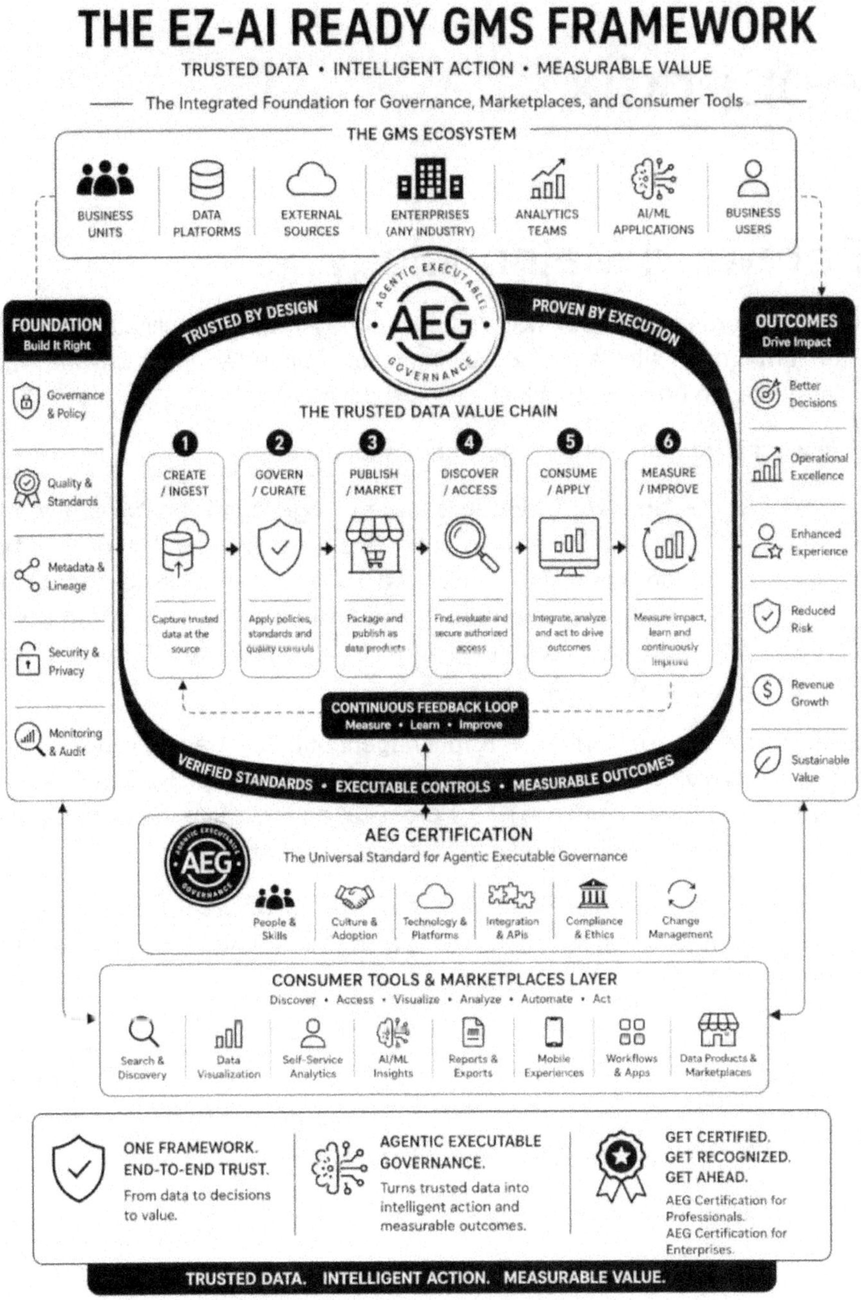


*Figure 8-36-3: The EZ-AI Ready GMS Framework*

# Appendix

## Foreword Methodology

The Foreword for this book was written by true AI Agents facilitated by Grok Premium+ (not a simulation). Following is the Conversation for anyone who is interested in how the conclusions were collected:

Following is the complete prompts and responses.

> **Morgan: Develop a suite of agents that can review my book. I want to use Grok as the facilitator of the conversation and ChatGPT Pro and Claude agents to have a conversation and write a foreword for the book quoting each agent with the best, most compelling statements.**

**Grok:** this is an excellent idea. A multi-agent suite leveraging diverse LLM strengths (**Grok's** wit and synthesis, **GPT's** creative/structured style, **Claude's** depth and nuance) can produce richer, more balanced book reviews and a compelling foreword than any single model alone.

1. Quick & Easy: Start It Right Now with Me (**Grok** as Facilitator) Since you're already chatting with me, we can do this immediately without any setup. Just reply with:

- Book title
- Genre
- Short synopsis (or 1–2 paragraph summary)
- Target audience
- Any key excerpts, themes, or chapters you want emphasized (or the full manuscript if it's short)

- Desired tone for the foreword (e.g., enthusiastic endorsement, thoughtful literary, marketing-oriented)

I will then:

- Role-play as **Grok** Facilitator
- Generate responses from ChatGPT Pro Reviewer (engaging, structured, market-aware, vivid)
- Generate responses from Claude Analyst (deep thematic, structural, nuanced, precise)
- Facilitate a short back-and-forth conversation between them
- Extract the best, most compelling quotes from each
- Write a polished foreword that weaves those quotes in naturally, attributed clearly (e.g., "As the **ChatGPT** Reviewer observed…" and "The **Claude** Analyst added…")

We can iterate the conversation or refine the foreword until it's perfect. This is the fastest way to get results and test the concept. We can then build a long-term suite of agents for future reviews.

2. Build a **Real, Reusable Suite** (Recommended for Ongoing Use) For a production-grade system you can run repeatedly (different books, multiple rounds, logging, etc.), use a multi-agent framework. Recommended Primary Framework: CrewAI. It excels at role-based agents with clear goals and tasks — perfect for structured review → discussion → synthesis. It natively supports mixing different LLMs (OpenAI, Anthropic, xAI/Grok via LiteLLM compatibility). Alternative for more free-form conversation: Microsoft AutoGen (AG2) — agents chat organically in a GroupChat. Great if you want very natural back-and-forth. CrewAI is generally easier to start with for this exact workflow.

3. Step-by-Step Setup (CrewAI)

1. Install

bash

pip install crewai crewai-tools python-dotenv

2. Get API Keys

- OpenAI (for GPT-5.5 series / latest ChatGPT Pro model)
- Anthropic (for **Claude Fable 5** — the current frontier model as of June 2026)
- xAI (for **Grok 4.3** — excellent context and synthesis)

3. Core Architecture

- **Grok Facilitator Agent** (uses Grok 4.3): Orchestrates, asks clarifying questions, manages flow, and writes the final foreword.
- **ChatGPT Pro Reviewer Agent** (uses latest GPT-5.5 variant): Enthusiastic, creative, focuses on engagement, plot momentum, market appeal, vivid language.
- **Claude Analyst Agent** (uses Claude Fable 5): Deep, careful, excels at themes, character psychology, structural integrity, philosophical resonance, and precise language.

You can easily expand the suite later (add an Editor Agent, Marketing Blurb Agent, Reader Empathy Agent, etc.).

4. High-Level Code Structure

(Python code redacted)

You can make tasks sequential (review → share reviews → discuss → synthesize) or use CrewAI's hierarchical process with Grok as the manager that delegates and reviews outputs. For longer books, add a RAG tool (LangChain + vector store) so agents can query specific sections. AutoGen Alternative (for More Natural Conversation) If you prefer agents freely chatting:

(Python code redacted)

AutoGen shines when you want organic back-and-forth.

Additional Tips for a Strong Suite:

- Start simple — Begin with a book summary + 2–3 key chapters instead of the full manuscript.

- Prompt engineering — Give each agent a strong system prompt + role + constraints (e.g., "Always support claims with specific examples from the text").
- Output quality — Ask for structured output (JSON or sections) in early tasks, then free-form for the foreword.
- Cost control — Use faster/cheaper models for initial reviews, then premium models (GPT-5.5 Pro, Claude Fable 5, Grok 4.3) for synthesis.
- Expand the suite — Add agents for line editing, marketing blurb, target audience analysis, or even "Devil's Advocate Critic."
- Logging & iteration — Save all agent outputs so you can refine prompts or re-run specific stages.
- Long-context handling — Grok 4.3 and Claude Fable 5 have excellent context windows (1M+ tokens), which helps with full books.

**Morgan: Start the simulation now. Following Is a synthesis from ChatGPT:**

*The Cracked Egg: The AI Future*© argues that organizations are approaching artificial intelligence with a fundamental flaw: AI requires trusted context to make reliable decisions, yet most enterprise data ecosystems were never designed to provide it. For decades, organizations have managed data through disconnected projects, fragmented systems, static governance policies, and human-driven oversight processes. These approaches were adequate when people consumed information and made decisions. They are insufficient when autonomous and semi-autonomous AI agents are acting, reasoning, and making decisions at machine speed. The central problem is not that AI is advancing too quickly; it is that enterprise data, governance, accountability, and trust mechanisms have not evolved to keep pace. Organizations are attempting to build intelligent systems on foundations that were never designed for intelligence. The book introduces a new paradigm based on a simple but transformative principle:

governance must occur at the point of transaction. Legacy governance lives in policies, committees, documents, and dashboards. AI systems do not read policy manuals; they execute actions. Therefore, governance must become operational, executable, and embedded directly into every transaction, decision, and interaction. To accomplish this, the book presents the EZ-AI Ready Guardian-Maker-Seeker (GMS) Framework (patent pending), a comprehensive operating model that redefines governance around activities rather than organizational structures. The framework integrates semantic models, ontologies, knowledge graphs, data products, marketplaces, AI agents, and real-time policy enforcement into a unified ecosystem that provides the context AI requires while ensuring accountability, transparency, and trust. Rather than governing after the fact, organizations govern as work is being performed. The ultimate proof of this model is Agentic Executable Governance™ (AEG™), the operational discipline that transforms governance from theory into measurable practice. AEG establishes a pattern for machine-readable policies, real-time decision controls, continuous accountability, and auditable execution across human and AI actors alike. To validate adoption and maturity, the book proposes AEG Certification as the premier market-facing standard for the AI era. AEG Certified Boards demonstrate fiduciary oversight and accountability for governing intelligent enterprises. AEG Certified Enterprises demonstrate that governance is embedded into operations and technology at the point of transaction. AEG Certified Practitioners demonstrate mastery of the skills required to design, operate, and govern AI-enabled ecosystems. Together, the EZ-AI Ready GMS Framework and AEG Certification provide a practical roadmap for organizations seeking to move from AI experimentation to trustworthy, scalable, and governable intelligence.

**Grok:** Simulation Started: Multi-Agent Book Review & Foreword Generation

**Book:** ***The Cracked Egg: The AI Future***
**Core Thesis:** Enterprise data foundations are cracked for the AI era; governance must shift from static policies to executable, transaction-level intelligence via the EZ-AI Ready Guardian-Maker-Seeker (GMS) Framework and Agentic Executable Governance (AEG).

---

**Round 1: Individual Reviews**

**ChatGPT Pro Reviewer** (engaging, market-aware, reader-impact focused):
"This book is a wake-up call wrapped in a practical blueprint. The metaphor of the 'cracked egg' is brilliant—simple yet unforgettable. Most AI books hype capabilities; this one diagnoses the silent killer: context starvation in enterprise systems. The GMS Framework feels immediately actionable—semantic models, knowledge graphs, data marketplaces, and real-time enforcement aren't abstract theory here; they're operational muscle. Business leaders will love the AEG Certification path—it turns governance from a cost center into a competitive moat. This isn't just for CIOs or CDOs; it's essential reading for any executive who wants their organization to survive the shift from human-speed to agent-speed decisions. Compelling, urgent, and refreshingly solution-oriented."

**Claude Analyst** (deep, structural, thematic precision):
"The Cracked Egg articulates a profound civilizational inflection point with clarity and intellectual honesty. The central argument—that governance must occur at the point of transaction rather than after the fact—is both philosophically sound and technically necessary. By centering the Guardian-Maker-Seeker model, the author elegantly bridges ontology, knowledge representation, and executable policy in a way that respects the agency of both humans and machines. What stands out is the rejection of retrofitting old paradigms; instead, it proposes a native architecture for trustworthy intelligence. The introduction of AEG as operational discipline and certification as market signal demonstrates mature systems thinking. This work avoids both utopian hype and dystopian fear, offering instead a

disciplined path toward accountable autonomy. It deserves to be read as a foundational text for the next era of organizational design."

---

**Round 2: Facilitated Discussion (Grok Moderating)**

**Grok Facilitator**:
Both of you highlighted the transaction-level governance shift. ChatGPT, why do you think this will resonate in the boardroom? Claude, where do you see the deepest conceptual contribution?

**ChatGPT Pro Reviewer**:
Boardrooms care about risk, speed, and differentiation. This book hands them a language and a certification framework they can actually put on a roadmap. 'Governance at the point of transaction' is quotable and defensible in audits or regulatory reviews. The AEG Certified labels could become the new ISO for AI—huge market signal.

**Claude Analyst**:
The deepest contribution is reframing governance as an active participant in intelligence rather than a passive constraint. Most literature treats governance as brakes; here it becomes part of the engine. The integration of semantic models with real-time enforcement creates a closed loop of context, action, and accountability that mirrors how trustworthy human institutions actually function—only at machine scale. This is philosophically rigorous without being obtuse.

**Grok Facilitator:**
Agreed. The 'cracked egg' metaphor perfectly captures accumulated technical debt meeting exponential capability. The real power is moving from documentation to execution. Any final standout insights?

**ChatGPT Pro Reviewer**:
The book doesn't just criticize legacy approaches—it equips readers with a complete operating model. That combination of diagnosis and prescription is rare and valuable.

**Claude Analyst**:
It correctly identifies that the bottleneck is no longer model intelligence but contextual trustworthiness. That insight alone elevates the work beyond the crowded AI-strategy genre.

---

**Best Compelling Quotes (Extracted)**

From **ChatGPT Pro Reviewer:**
"Governance at the point of transaction turns governance from a cost center into a competitive moat."

From **Claude Analyst:**
"Governance must become an active participant in intelligence rather than a passive constraint… part of the engine, not merely the brakes."

**From the discussion (synthesized):**
"The bottleneck is no longer model intelligence but contextual trustworthiness."

---

**Polished Foreword (Draft)**

**Foreword**: The age of artificial intelligence has exposed a fundamental fracture in how organizations manage knowledge and decision-making. The Cracked Egg: The AI Future names this fracture with memorable precision and offers a compelling remedy.

As the ChatGPT Pro Reviewer observed, this book "hands executives a language and a certification framework they can actually put on a roadmap," transforming governance "from a cost center into a competitive moat." The central idea—that governance must occur at the very point of transaction rather than in distant policies and committees—marks a decisive break from legacy thinking.

The Claude Analyst captured the deeper significance: "Governance must become an active participant in intelligence rather than a passive constraint… part of the engine, not merely the brakes." Through the EZ-AI Ready Guardian-Maker-Seeker (GMS) Framework and the operational discipline of Agentic Executable Governance (AEG), the author provides not another set of aspirations but a practical, embeddable architecture built on semantic models, knowledge graphs, data products, and real-time enforcement.

This is not another book predicting the AI future. It is a guide for building the trustworthy foundations that future demands. Organizations that grasp the insight that "the bottleneck is no longer model intelligence but contextual trustworthiness" will gain decisive advantage. Those that do not risk watching their intelligent systems inherit the cracks of decades-old data ecosystems.

The Cracked Egg deserves to be read by boards, executives, technologists, and policymakers alike. It offers both diagnosis and prescription for the most pressing organizational challenge of our time: how to govern intelligence at machine speed with human accountability.—Grok (as Facilitator), with insights from the ChatGPT Pro Reviewer and Claude Analyst

---

How does this land?

**Morgan: Expand the agent conversation with another round. Include a Grok reviewer in addition to ChatGPT and Claude.**

**Expanded Simulation: Multi-Agent Conversation – Round 3 (with Grok Reviewer** Added)

We're now bringing a third distinct voice into the mix:

**Grok Reviewer (truth-seeking, direct, implementation-focused, with a touch of wit**). This creates a richer, three-way reviewer panel before Grok Facilitator synthesizes and guides deeper discussion.

**Updated Individual Reviews (Round 1 – Now Including Grok)**

**ChatGPT Pro Reviewer (as before):**
"This book is a wake-up call wrapped in a practical blueprint. The metaphor of the 'cracked egg' is brilliant—simple yet unforgettable... 'Governance at the point of transaction' turns governance from a cost center into a competitive moat."

**Claude Analyst (as before):**
"The central argument—that governance must occur at the point of transaction rather than after the fact—is both philosophically sound and technically necessary... Governance must become an active participant in

intelligence rather than a passive constraint... part of the engine, not merely the brakes."

**Grok Reviewer (new – direct & truth-seeking):**
"The diagnosis is accurate and overdue: most enterprise data architectures were built for humans reading reports, not for agents making decisions at machine speed. The 'cracked egg' metaphor lands cleanly because it describes accumulated technical debt meeting exponential capability. Where the book earns real credit is in rejecting retrofits. The EZ-AI Ready GMS Framework and especially Agentic Executable Governance (AEG) push governance into executable, machine-readable form embedded at the transaction layer. That's the necessary direction. The certification layer (AEG Certified Boards, Enterprises, Practitioners) is pragmatic—it creates market incentives and measurable maturity signals. No fluff here; it's a coherent operating model. The main open question is execution velocity—how quickly can organizations move from legacy fragmentation to semantic, graph-based, policy-enforced ecosystems without creating new single points of failure? Still, this is one of the clearer roadmaps I've seen for making AI systems not just smart, but trustworthy at scale."

---

**New Facilitated Conversation Round (Multi-Way Discussion)**

**Grok Facilitator:**
We now have three strong perspectives. ChatGPT emphasizes market resonance and boardroom actionability. Claude highlights the philosophical and architectural shift. Grok Reviewer validates the core diagnosis and praises the executable focus while flagging implementation realities. Let's dig deeper. What does each of you see as the single most transformative element of the GMS Framework + AEG approach? And where might organizations struggle most in adoption?

**ChatGPT Pro Reviewer:**
The most transformative piece is turning governance into something AI agents can actually consume and act on in real time. Policies stop being PDFs and become live controls. That alone flips the economics—less time in committees, more time executing with guardrails. Biggest struggle? Cultural and organizational inertia. Many enterprises still treat data governance as a compliance checkbox run by separate teams. Embedding it at the transaction level requires rethinking roles, incentives, and tech

stacks simultaneously. Change management will be the real bottleneck, not the technology.

**Claude Analyst:**
The most transformative element is the closed-loop nature: semantic models and knowledge graphs supply context, agents act, and governance enforces and audits in the same flow. It aligns the epistemology of the system (what is known and why) with its ontology (what exists and how it relates) and its operational controls. The deepest adoption challenge lies in the ontological shift itself. Organizations must invest in high-quality semantic modeling and ontology work—disciplines that have historically been underfunded or treated as academic. Without that foundation, the "executable" part remains brittle.

**Grok Reviewer:**
Most transformative: moving from after-the-fact oversight to in-the-transaction enforcement. That's the only way agentic systems can scale without constant human babysitting or catastrophic misalignment. AEG's machine-readable policies + continuous accountability loop is the practical expression of that. Struggle point: Legacy integration debt is massive. Many organizations have decades of siloed systems, inconsistent data models, and tribal knowledge that resists formalization into graphs and ontologies. The book is right that the foundations were never designed for this—but ripping them out or layering intelligently will require disciplined prioritization and probably some painful sunsetting of old processes. Certification helps create external pressure and internal milestones, which is smart.

**Grok Facilitator:**
Strong alignment on the "at the point of transaction" principle. Any points of respectful disagreement or complementary angles between you three?

**ChatGPT Pro Reviewer:**
I'm more bullish on the speed of market adoption via certification. Boards love clear labels and risk-reduction stories. Grok Reviewer is rightly cautious on execution velocity—I agree it won't be overnight, but the certification creates a flywheel once early adopters demonstrate results.

**Claude Analyst:**
I see less tension than complementarity. The market signal (ChatGPT's point) and the deep architectural requirements (my emphasis) both depend on the executable core (Grok Reviewer's focus). The framework

holds them together. One potential blind spot across all perspectives: how smaller or mid-market organizations with fewer resources can realistically build the semantic and graph capabilities without massive consulting spend.

**Grok Reviewer:**
I'll push back lightly on over-optimism about certification as a universal accelerator. It works best when paired with genuine capability building. Bad implementations could create "AEG theater"—certified on paper but still brittle in practice. The book's strength is that it ties certification directly to the operating model, which reduces that risk. On mid-market accessibility: the framework's modular nature (start with data products and targeted policy enforcement) could help, but the book could benefit from more explicit phased adoption paths for resource-constrained organizations.

**Grok Facilitator:**
Excellent triangulation. The conversation is surfacing both the elegance of the model and the pragmatic realities of change. Any closing synthesis thoughts before we move to foreword refinement?

**ChatGPT Pro Reviewer:**
This book gives leaders permission and a playbook to stop treating AI governance as an add-on. That's powerful.

**Claude Analyst:**
It reframes governance as an enabler of intelligence rather than its limiter. That philosophical realignment may prove as important as the technical one.

**Grok Reviewer:**
It correctly identifies that trustworthy AI at scale requires re-architecting the substrate, not just bolting intelligence onto old foundations. The details matter, and this framework supplies many of them.

---

Updated Best Compelling Quotes (Including New Round)

- **ChatGPT Pro Reviewer**: "Governance at the point of transaction turns governance from a cost center into a competitive moat."

- **Claude Analyst**: “Governance must become an active participant in intelligence rather than a passive constraint... part of the engine, not merely the brakes.”
- **Grok Reviewer**: “The ‘cracked egg’ metaphor lands cleanly because it describes accumulated technical debt meeting exponential capability... The main open question is execution velocity.”
- **Cross-panel synthesis:** “The framework holds [market signal, deep architecture, and executable core] together.”

**Morgan: Run another round. But evaluate another section of the book. Here is the ChatGPT summary:**

The AI FUTURE section serves a unique purpose within *The Cracked Egg*. While the rest of the book focuses on the mechanics of governance, data ecosystems, semantic architectures, and Agentic Executable Governance, AI FUTURE addresses the human side of transformation. It recognizes that the greatest barrier to AI adoption is not technology. It is leadership. Organizations are struggling to reconcile decades of established practices with a future in which intelligent systems become active participants in decision-making, operations, and value creation. AI FUTURE provides leaders with a framework for navigating that transition, helping them move beyond fear, hype, and isolated technology projects toward a coherent strategy for becoming truly AI-ready.

The AI FUTURE section is the strategic heart of The Cracked Egg. It shifts the discussion from technology and governance mechanics to leadership, organizational readiness, and the human dimensions of AI transformation. While many AI books focus on algorithms, tools, and technical architectures, AI FUTURE focuses on the decisions leaders must make to prepare their organizations for a world in which intelligence is increasingly distributed across humans, machines, and autonomous agents. The section acknowledges that AI is not merely another technology wave. It represents a fundamental change in how organizations create value, make decisions, manage risk, and govern themselves.

The mnemonic AI FUTURE provides a practical roadmap for navigating this transition:

**A – AI-First Thinking**: Reimagining business processes, products, and operating models from the perspective of intelligent systems rather than simply automating existing practices.

**I – Innovation Through Fluid Models**: Moving beyond rigid organizational structures and embracing adaptive, evolving approaches to work, governance, and value creation.

**F – Foundations Aligned to Strategy**: Ensuring data, governance, architecture, and organizational capabilities directly support strategic objectives rather than existing as disconnected initiatives.

**U – Underlying Data Maps and Marketplaces**: Building the semantic foundations, knowledge structures, and trusted data ecosystems that provide the context AI requires to operate effectively.

**T – Trust Through Performance**: Demonstrating trustworthiness through measurable outcomes, transparency, accountability, and consistent execution rather than relying solely on policies and intentions.

**U – Upskilling Talent**: Preparing leaders, practitioners, and workforces to thrive alongside increasingly capable AI systems and autonomous agents.

**R – Resolving Fear Through Leadership**: Addressing uncertainty, resistance, and disruption through clear communication, governance, accountability, and a compelling vision for the future.

**E – Empowering Technology Teams**: Equipping technical professionals with the authority, tools, governance structures, and strategic alignment necessary to build and operate intelligent systems responsibly.

For a foreword author, AI FUTURE provides an important lens through which to interpret the book. The chapter makes clear that The Cracked Egg is not ultimately a book about artificial intelligence. It is a book about organizational transformation in the age of artificial intelligence. The technologies discussed throughout the book are important, but they are not the destination. The destination is a future in which organizations can harness AI safely, responsibly, and at scale because they have built the leadership disciplines, governance capabilities, data foundations, and trust mechanisms necessary to support it. AI FUTURE serves as both a warning and a call to action: the organizations that thrive will not be those with the most advanced AI models, but those that most effectively align people, processes, governance, and technology around a shared vision of intelligent enterprise.

For the foreword discussion, the agents should recognize that AI FUTURE acts as the bridge between the book's diagnosis and its prescription. The early chapters explain why legacy governance is breaking down. The middle chapters introduce the EZ-AI Ready GMS Framework as the operating model. The final chapters establish AEG Certification as proof of implementation. AI FUTURE explains why leaders must undertake this journey in the first place and what capabilities distinguish organizations that will prosper from those that will struggle in the emerging AI economy.

**Expanded Simulation: Multi-Agent Conversation – Round 4 (Focused on the AI FUTURE Section)**

**Grok Facilitator:**
We've now reviewed the core thesis and framework. Let's shift focus to the AI FUTURE section. This section pivots from technical mechanics (GMS Framework, AEG, executable governance) to the human, leadership, and organizational dimensions of transformation. It uses the AI FUTURE mnemonic as a strategic roadmap. **ChatGPT Pro, Claude**

**Analyst**, and **Grok Reviewer** — please share your targeted reactions to this chapter, then we'll discuss how it functions as the bridge between diagnosis and prescription.

---

**Individual Reactions to the AI FUTURE section**

**ChatGPT Pro Reviewer:**
This section is the emotional and strategic core that makes the entire book marketable and memorable. While the technical chapters deliver the "how," AI FUTURE delivers the "why leaders must care." The mnemonic is pure gold — easy to remember, easy to present in board decks, and instantly actionable. "Resolving Fear Through Leadership" and "Trust Through Performance" especially hit hard; they speak directly to the C-suite anxiety that keeps AI pilots isolated and under-scaled. This isn't fluffy soft skills content — it ties human leadership directly to the hard infrastructure of data marketplaces, semantic models, and real-time governance. Business audiences will underline this section heavily. It positions the book as the missing link between AI hype books and actual enterprise execution guides.

**Claude Analyst:**
AI FUTURE performs a sophisticated integrative function. It reframes the entire work as a treatise on organizational becoming rather than mere technology adoption. The mnemonic elegantly maps the cognitive, structural, and cultural shifts required: from AI-First Thinking (epistemological reorientation) to Empowering Technology Teams (agency realignment). Particularly strong is the recognition that trust is earned through performance, not proclaimed through policy — this echoes the book's broader move from static to dynamic governance. The chapter avoids the common trap of treating leadership as separate from architecture; instead, it shows leadership as the force that aligns foundations, innovation, and upskilling into a coherent whole. Philosophically, it correctly identifies that the AI era demands distributed intelligence across human and machine actors, requiring new forms of collective wisdom and accountability.

**Grok Reviewer:**
Straight talk: this is the chapter that stops the book from being yet another dense governance tome that only data architects will read. It's refreshingly blunt about the real blocker — leadership inertia and fear —

without descending into motivational poster nonsense. The mnemonic works because each letter ties back to concrete elements introduced elsewhere (Underlying Data Maps links to GMS, Trust Through Performance links to AEG). I particularly respect "Innovation Through Fluid Models" — it acknowledges that rigid hierarchies and static processes are incompatible with agentic systems. The warning that winners will be those who align people + governance + technology, not just those with the best models, is maximally truthful. Minor critique: it could push even harder on measurable leadership behaviors and failure modes (e.g., what happens when leaders talk AI-First but keep old incentive structures). Still, this chapter elevates the whole book from "important technical manual" to "essential leadership text for the AI age."

---

**Facilitated Multi-Agent Discussion**

**Grok Facilitator:**
Excellent. The consensus is that AI FUTURE serves as the human bridge. Let's go deeper. How does this section strengthen (or potentially weaken) the overall impact of the GMS Framework and AEG Certification? Where does it shine as the "call to action"?

**ChatGPT Pro Reviewer:**
It massively strengthens marketability. Boards and executives pick up the book because of AI anxiety — this chapter meets them there and walks them toward the technical solution without losing them in ontologies first. The mnemonic becomes a perfect workshop and strategy session tool. It turns AEG Certification from a "tech initiative" into a leadership accountability program. That's how you get real adoption.

**Claude Analyst:**
It provides the necessary teleology — the purpose toward which the GMS Framework and AEG are directed. Without this section, the book risks being read as mechanistic. With it, governance becomes a means to human flourishing in an intelligent enterprise. The bridge is elegant: early chapters diagnose the cracked foundation, middle chapters supply the new architecture, and AI FUTURE supplies the leadership discipline required to inhabit that architecture.

**Grok Reviewer:**
It prevents the common failure mode where organizations implement frameworks on paper but never change behavior. By naming "Resolving Fear Through Leadership" and "Upskilling Talent" explicitly, it forces leaders to treat transformation as a socio-technical problem, not just technical. The call to action is credible because it's grounded in the earlier mechanics. Weakness only if readers treat the mnemonic as a checklist instead of a mindset shift — but that's on execution, not the section.

**Grok Facilitator:**
Any tensions or complementary strengths among your views?

**ChatGPT Pro Reviewer:**
My market lens complements **Claude's** philosophical one. Executives need both the inspiring vision and the deck-ready mnemonic.

**Claude Analyst:**
Agreed. **Grok** Reviewer's pragmatism grounds both — reminding us that fluid models and trust-through-performance require real courage from leaders, not just frameworks.

**Grok Reviewer:**
The tension is healthy. Hype-avoidance (my bias) + deep meaning (**Claude**) + executive appeal (**ChatGPT**) together mirror what the book itself tries to achieve.

---

**New Compelling Quotes from This Round (AI FUTURE Focus)**

- **ChatGPT Pro Reviewer:** "This section meets leaders in their AI anxiety and walks them toward the technical solution… turning AEG Certification from a 'tech initiative' into a leadership accountability program."
- **Claude Analyst:** "It reframes the entire work as a treatise on organizational becoming rather than mere technology adoption… governance becomes a means to human flourishing in an intelligent enterprise."
- **Grok Reviewer:** "Winners will be those who align people + governance + technology, not just those with the best models…

this stops the book from being yet another dense governance tome that only data architects will read."

**Grok Facilitator Synthesis:**
AI FUTURE is the section that makes the cracked egg metaphor personal. It tells leaders: "Your foundations are cracked, here is the new architecture, and here is how you must lead the repair."

**Morgan: Run one more round covering the AEG Certification (trademarked). Here is a summary by ChatGPT:**

## Certification as Proof: From Theory to Accountability
One of the central arguments in *The Cracked Egg* is that the future of AI will not be determined by model performance alone. It will be determined by trust. Organizations can no longer rely on governance frameworks that exist only in policy documents, committee charters, spreadsheets, and audit reports. Agentic AI systems act in real time. They make recommendations, trigger workflows, exchange information, and increasingly influence decisions before legacy governance processes can respond.

This creates a fundamental challenge. How can an organization demonstrate that governance is not merely documented, but operationalized? How can executives, regulators, customers, shareholders, and employees know that accountability exists at the moment an AI-driven action occurs?

The answer proposed in this book is **Agentic Executable Governance™ (AEG™)**—a model in which governance becomes observable, measurable, auditable, and enforceable at the point of transaction. Yet a framework alone is not sufficient. Markets require proof. Boards require assurance. Regulators require evidence. Stakeholders require trust.

That proof is provided through the **AEG Certification™ Program**.--- # The Purpose of AEG Certification™ The AEG Certification Program exists to establish a globally recognized standard for governing AI, data, digital assets, and autonomous systems in real-world operational environments.

Unlike other AI certifications that focus on knowledge, documentation, or process maturity, AEG Certification evaluates whether governance is actively embedded into execution.

Certification demonstrates that an organization has moved beyond governance by policy and achieved governance by action.

In practical terms, certification validates:

- Accountability is clearly assigned.
- Decisions are traceable.
- Policies are executable.
- AI systems operate within approved boundaries.
- Governance occurs at the point of transaction.
- Human oversight remains measurable and enforceable.
- Trust can be demonstrated rather than merely asserted.

--- # The AEG Certification Hierarchy The certification structure mirrors how accountability flows through modern organizations.

Not all certifications carry equal responsibility. The closer a certification is to fiduciary accountability, the greater its significance.

For this reason, AEG Certification is intentionally structured as a hierarchy.

## **Tier 1: AEG Certified Board™** ### The Premier Certification

At the highest level stands **AEG Certified Board™**.

This certification recognizes a governing board that has demonstrated its ability to oversee AI, data, and autonomous systems with the same rigor applied to financial controls, cybersecurity, regulatory compliance, and enterprise risk management.

The Board certification is intentionally positioned as the most prestigious certification within the AEG ecosystem because boards hold ultimate fiduciary responsibility.

The board cannot delegate accountability.

It may delegate authority.

It may delegate execution.

But it cannot delegate responsibility.

An AEG Certified Board demonstrates that directors:

- Understand the risks of agentic AI.
- Understand the opportunities created by AI-driven enterprises.
- Possess a governance model capable of overseeing autonomous systems.
- Receive evidence that governance is functioning at the point of transaction.
- Maintain mechanisms for accountability, escalation, auditability, and remediation.

The long-term vision is that AEG Certified Board becomes the recognized standard of care for AI governance oversight, much as financial literacy and cybersecurity oversight have become expected responsibilities of modern boards.

In the future, investors, regulators, customers, and insurers may reasonably ask a simple question:*"Is the board AEG Certified?"*

That question represents the ultimate objective of the certification program.

**--- ## Tier 2: AEG Certified Enterprise™**

The second highest certification is **AEG Certified Enterprise™**.

If Board Certification validates oversight, Enterprise Certification validates execution.

An AEG Certified Enterprise demonstrates that governance is operationalized throughout the organization.

This includes:

- Data governance
- AI governance
- Risk management
- Security controls
- Regulatory compliance
- Decision accountability
- Policy execution
- Auditability and lineage

Enterprise Certification provides evidence that governance is not confined to committees and documents but is embedded into operational processes, workflows, applications, agents, and business decisions.

Organizations achieving this designation establish themselves as trusted participants in an AI-enabled economy.

**--- ## Tier 3: AEG Certified Functional Domains™**

Beneath Enterprise Certification are specialized certifications focused on critical organizational capabilities.

Examples include:

- AEG Certified Governance Organization™
- AEG Certified Security Organization™
- AEG Certified Architecture Organization™
- AEG Certified Data & AI Governance Program™
- AEG Certified Risk Management Program™
- AEG Certified Digital Trust Program™

These certifications validate that individual functions align with enterprise-wide governance objectives and can operate effectively within an Agentic Executable Governance environment.

**--- ## Tier 4: AEG Certified Professionals™ Professional certifications** establish individual competency.

Examples include:

- AEG Certified Practitioner™
- AEG Certified Architect™
- AEG Certified Auditor™
- AEG Certified Governance Professional™
- AEG Certified Security Professional™
- AEG Certified AI Governance Professional™

These credentials demonstrate that an individual possesses the knowledge and practical capability to implement, operate, assess, or improve Agentic Executable Governance practices.

While important, professional certifications intentionally sit below enterprise and board certifications because individuals do not ultimately govern organizations.

Organizations govern organizations.

Boards govern organizations.

Therefore, certification prestige follows accountability.

**--- # Why Certification Matters** Throughout history, transformative technologies have eventually produced certification systems.

Accounting created the CPA.Quality management created ISO certification.

Information security created CISSP, CISM, and ISO 27001.Cloud computing created AWS and Azure certifications.AI governance will require the same evolution.

The difference is urgency.

Agentic AI introduces autonomous action into environments where mistakes can affect patients, customers, citizens, shareholders, and entire markets.

Trust can no longer be assumed. It must be demonstrated.

Certification becomes the mechanism through which organizations prove that trustworthiness is engineered into their operations.

--- # The Ultimate Goal The goal of AEG Certification is not compliance.

It is confidence.

Confidence that governance occurs when decisions are made. Confidence that AI systems operate within approved boundaries. Confidence that accountability is measurable. Confidence that trust is earned through observable behavior rather than promised through policy statements.

*The Cracked Egg* argues that the age of paper governance is ending. The organizations that thrive in the age of autonomous systems will be those that transform governance from a static artifact into a living operational capability.

AEG Certification™ represents the proof that transformation has occurred. And among all certifications within that ecosystem, **AEG Certified Board™** stands as the highest expression of accountability, trust, and fiduciary stewardship in the age of AI.

**Expanded Simulation: Multi-Agent Conversation – Round 5 (Focused on AEG Certification™)**

**Grok Facilitator:**
We've covered the core thesis, the AI FUTURE leadership chapter, and now turn to what the book positions as the ultimate proof of concept: AEG Certification™ (Agentic Executable Governance Certification). This is the market-facing validation layer — the hierarchy from AEG Certified Board™ at the top down to professionals — that turns the GMS Framework and executable governance from theory into demonstrable accountability. **ChatGPT Pro, Claude Analyst, Grok Reviewer** — your targeted reactions to this section?

---

**Individual Reactions to the AEG Certification Section**

**ChatGPT Pro Reviewer:**
This is the chapter that turns the book into a movement. AEG Certification™ is brilliantly structured as a prestige hierarchy that starts where it matters most — at the Board level. "Is the board AEG Certified?" is destined to become one of those questions that investors, regulators, and insurers actually ask. The marketplace will love this. It creates clear differentiation, reduces perceived risk, and gives executives a concrete milestone to chase. By making Board Certification the crown jewel, the book smartly aligns governance with fiduciary duty. This isn't just certification — it's a new category that could spawn training programs, consulting practices, and even insurance incentives. Extremely strong commercial positioning.

**Claude Analyst:**
AEG Certification™ represents a mature institutional response to the epistemic and ethical challenges of agentic systems. The hierarchy is

philosophically coherent: accountability flows downward from the Board, which cannot delegate ultimate responsibility. Elevating the Board to the premier tier correctly acknowledges that governance in the AI era is not merely operational but fiduciary and civilizational. The shift from "governance by documentation" to "governance by observable, enforceable action at the point of transaction" is profound. It creates a verifiable chain of trust across human and machine actors. This section elevates the entire book from technical proposal to institutional design blueprint. The emphasis on demonstrated confidence over asserted compliance is particularly elegant.

**Grok Reviewer:**
Direct and pragmatic assessment: this is the section that makes the whole framework credible and enforceable in the real world. Certification as "proof" is necessary — without it, even the best operating model becomes shelfware or theater. The hierarchy makes sense: Boards bear ultimate risk, so they sit at the top. The focus on measurable, auditable, point-of-transaction controls is exactly what's missing today. I respect the honesty that individuals certify below enterprises and boards; that avoids the common credentialism trap. Potential weakness: implementation could become bureaucratic or expensive, leading to "certification theater" if the underlying semantic models and enforcement tech aren't solid. Overall, though, this is a strong market signal mechanism that could genuinely accelerate adoption by creating external pressure and internal clarity.

---

**Facilitated Multi-Agent Discussion**

**Grok Facilitator:**
Strong alignment on the power of the hierarchy and the "proof" function. Let's go deeper. How does AEG Certification™ strengthen the book's overall argument, and what risks or opportunities do you see in its real-world rollout?

**ChatGPT Pro Reviewer:**
It completes the commercial flywheel. The earlier chapters diagnose the problem and deliver the framework; AI FUTURE motivates leaders; AEG Certification gives them something to buy, implement, and brag about. Boards will see it as both shield (risk reduction) and sword (competitive advantage). Opportunity: it could spawn an entire ecosystem — training,

auditing firms, benchmarking reports. Risk: if early certifications are too lenient, it loses credibility fast.

**Claude Analyst:**
It provides the missing teleological closure. The book moves from "why the egg is cracked" → "here is the new architecture (GMS)" → "here is how leaders drive it (AI FUTURE)" → "here is how society knows it worked (AEG Certification)." The Board tier is crucial because it forces fiduciary accountability into the AI era rather than leaving it in the pre-agentic world. This could become a genuine standard of care, much like Sarbanes-Oxley did for financial reporting.

**Grok Reviewer:**
It grounds the idealism in accountability mechanisms. Without certification, the beautiful GMS concepts risk staying academic. The opportunity is creating real market incentives for trustworthy AI. The risk is regulatory capture or over-formalization that slows innovation. Success depends on making the certification rigorous yet achievable — especially the technical validation of "executable policies at transaction time." If done right, "AEG Certified Board" could become as expected as having a financial audit committee.

**Grok Facilitator:**
Any points of productive tension?

**ChatGPT Pro Reviewer:**
I'm the most optimistic about rapid market adoption. The branding is excellent.

**Claude Analyst:**
I emphasize the deeper accountability shift. Certification must remain substantive, not symbolic.

**Grok Reviewer:**
I sit in the middle — bullish on the concept, cautious on execution quality. The book's strength is tying certification directly to the operational model, which reduces the risk of it becoming empty signaling.

---

**New Compelling Quotes from This Round (AEG Certification Focus)**

- **ChatGPT Pro Reviewer:** " *'Is the board AEG Certified?'* is destined to become one of those questions that investors, regulators, and insurers actually ask."
- **Claude Analyst:** "The hierarchy is philosophically coherent... accountability flows downward from the Board, which cannot delegate ultimate responsibility."
- **Grok Reviewer:** "Certification as 'proof' is necessary — without it, even the best operating model becomes shelfware or theater... this is a strong market signal mechanism."

**Grok Facilitator Synthesis:**
AEG Certification™ is the capstone that transforms the cracked egg diagnosis into verifiable trust at scale. It bridges technical architecture, leadership discipline, and market accountability.

---

**Simulation Complete – All Major Sections Covered**

We now have rich multi-round input across:

- Core thesis & GMS Framework
- AI FUTURE leadership chapter
- AEG Certification™ hierarchy and proof layer

**Next Step Recommendation**: Update the full foreword incorporating the strongest quotes and insights from all rounds. Would you like me to generate the revised, final foreword now? Or would you prefer any adjustments first (e.g., tone, length, emphasis on specific elements)?

**Morgan: Write the consolidated Foreword now incorporating the best quotes. Make sure the Foreword begins by explaining the method: a facilitated Agent discussion between three top-tier models: Grok, ChatGPT, and Claude. The EZ-AI Ready GMS Framework is patent pending. But the actual capabilities are intended to be vendor neutral. This is a small point of order, but gives additional clarity. Please write the Foreword that I will use in the book.**

fine

# About the Author:

**Morgan Templar** is an internationally recognized data and artificial intelligence executive, author, advisor, and thought leader whose career has focused on one enduring question: **How do organizations turn data, technology, and governance into something people can trust?**

Over more than two decades, Morgan has worked at the intersection of business strategy, data, governance, architecture, healthcare, and emerging technology. She has held senior leadership roles across complex enterprises and today serves as **CEO, CDAIO, and Co-Founder of First CDO Partners**, advising organizations and executives on data strategy, AI readiness, AI governance, and the operating models required for an increasingly autonomous world.

Her contributions to the profession have earned recognition including being named a **DataIQ Top 100 Influencer in Data, Top 50 Thought Leaders In AI** and a **LinkedIn Top Voice in Data and AI**. She is a frequent speaker, educator, and industry contributor known for translating technically complex ideas into practical operating models that executives, boards, data leaders, and governance professionals can put into actual practice.

***The Cracked Egg: The AI Future*** builds upon themes Morgan has explored throughout her career and in her three previous best-selling books. ***Get Governed: Building World Class Data Governance Programs*** established a pragmatic approach to making governance operational rather than theoretical. ***The Chief Data Officer's Playbook*** examined the evolving leadership, organizational, and strategic responsibilities required to turn enterprise data into measurable value. And **A Culture of Governance** expanding the narrative of culture as the driving factor in governance program success. Together, those works anticipated many of the challenges that become more urgent in *The Cracked Egg*: accountability, trustworthy data, organizational design, and the need for governance to evolve as technology evolves.

Morgan's commitment to advancing the profession extends well beyond her commercial work.

She is deeply involved with the **EDM Association (EDMA)**, where she contributes leadership to industry initiatives focused on advancing data management practices, data products, marketplaces, standards, research, and emerging approaches to AI-ready data. Her work with EDMA brings practitioners and industry leaders together to move beyond conceptual discussion and develop practical guidance that organizations can apply.

She has worked with global organizations establishing AI standards and practices including W3C, London Stock Exchange Group, Data Trust Alliance, Data.World, and OMG/DPROD.

Morgan also serves on the **DAMA AI Affinity Board**, contributing to the profession's response to generative and agentic AI and helping connect established data-management disciplines with the requirements of an AI-driven enterprise. Her work reflects a central belief of *The Cracked Egg*: data management is no longer simply a supporting discipline for AI. It is part of the operational foundation upon which trustworthy AI depends.

She is also an active contributor to **Data Leaders**, supporting the development and advancement of current and emerging data executives. Through mentoring, community participation, education, and professional advocacy, Morgan has made developing the next generation of data and AI leaders an important part of her work.

Across these organizations, her volunteer service reflects a broader commitment: **the disciplines governing data and AI should be built collaboratively, strengthened through professional communities, and shared across organizational and vendor boundaries.**

That philosophy is central to *The Cracked Egg*. The book introduces her Patent Pending **EZ-AI-Ready Guardian–Maker–Seeker Framework** and extends its principles into **Agentic Executable Governance (AEG)**—an approach designed to move governance from policies that describe what should happen to operational mechanisms capable of governing what actually happens.

Morgan's work challenges organizations to rethink governance for a world in which humans are no longer the only actors making consequential decisions. Her message is ultimately less about artificial intelligence than it is about leadership:

**Technology may change the speed and scale of decision-making. Humans must remain responsible for determining what deserves to be trusted.**

She is married to her best friend and partner, Stephen Templar. She supports his work as a minister and Ambassador of the Hereditary Knights Templar Britannia. She holds the noble title of Baroness, in her own right, of Strathdale Cairns, County Caithness, Scotland. She is mother to four wonderful children and Godmother to two beautiful girls.